P9-CMS-477

Building Acoustics

Building Acoustics

Edited by

B. F. DAY, R. D. FORD
and P. LORD

University of Salford

ELSEVIER PUBLISHING COMPANY LIMITED
AMSTERDAM — LONDON — NEW YORK
1969

ELSEVIER PUBLISHING COMPANY LTD.
BARKING, ESSEX, ENGLAND

ELSEVIER PUBLISHING COMPANY
335 JAN VAN GALENSTRAAT, P.O. BOX 211, AMSTERDAM
THE NETHERLANDS

AMERICAN ELSEVIER PUBLISHING COMPANY INC.
52 VANDERBILT AVENUE, NEW YORK, N.Y. 10017

444-20047-9

LIBRARY OF CONGRESS CATALOG CARD NUMBER 73-95655

WITH 70 ILLUSTRATIONS AND 16 TABLES

Printed in Great Britain by Page Bros. (Norwich) Ltd., Norwich

Contents

CONTRIBUTING AUTHORS

B. F. Day, M.Sc., A.Inst.P., University of Salford

R. D. Ford, B.Sc.(Eng.), Ph.D., A.Inst.P., University of Salford

T. I. Hempstock, B.Sc., Ph.D., University of Salford

Ir.P.A. de Lange, T.P.D., T.N.O.enT.H., Delft*

A. B. Lewis, B.Sc., A.Inst.P., University of Salford

P. Lord, B.Sc., M.Sc.Tech., Ph.D., F.Inst.P., M.I.E.E., University of Salford

D. J. Saunders, B.Sc., Ph.D., A.Inst.P., University of Salford

A. W. Walker, B.Sc., Ph.D., Grad.Inst.P., University of Salford†

* Ir.P.A. de Lange is now Professor of Building Technology at the University of Eindhoven.

† Dr. A. W. Walker now works for Acoustical Investigation and Research Organisation Limited, Hemel Hempstead, Hertfordshire.

FOREWORD

This book is based on a series of lectures given by the staff of the Acoustics Group of the University of Salford during a one-week course, which at the time of writing has been repeated over a period of several years. The course is aimed quite specifically at those in the building industry and in architecture who need better understanding of how to control the acoustic environment.

It should be clearly understood that this is not a highly erudite textbook of acoustics, but is an attempt to present to the two professions of building and architecture a concise account of the basic principles involved in, for example, the techniques of sound insulation, the propagation of noise from motorways, and sound absorption. Much has been written elsewhere on the subject, but generally by scientists and engineers who expect the reader to have a reasonable knowledge of mathematics and go to considerable lengths to show how various relationships, for instance between superficial mass and sound insulation, are derived. In contrast the philosophy behind this production is to assume that most readers merely wish to see the information presented in either graphical or tabular form with perhaps only qualitative reasons given to justify it. Naturally there are those who will find this treatment inadequate and will wish to delve further, in which case they are recommended to read some of the publications referred to at the end of the book.

The authors are greatly indebted to the students who have attended their courses and through criticism, suggestions and interest have provided the incentive to produce this little book.

B. F. Day
R. D. Ford
P. Lord

CHAPTER 1

Basic Acoustics

1.1 Introduction

This is an elementary account of the fundamental concepts of acoustics and methods and units of measurement are discussed in the light of the present approach to investigating practical noise problems.

1.2 Sound Sources and Vibrations

Sound may be defined in general as the transmission of energy through solid, liquid or gaseous media in the form of vibrations. An alternative definition is that sound is the sensation produced through the stimulation of the ear and brain resulting from variations in the pressure of the air. These variations consist of alternate compressions and rarefactions or decompressions superimposed on the constant atmospheric pressure as shown in Figure 1.1. They are the result of

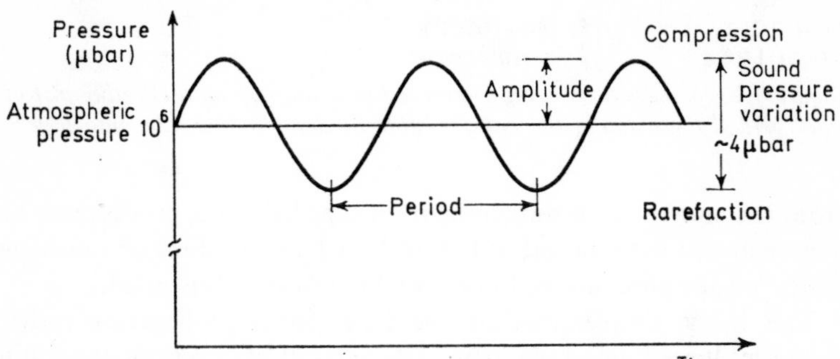

Figure 1.1. Variation in atmospheric pressure due to some vibrating object.

the vibrations of air molecules about a mean position which are in turn commonly excited by some vibrating object.

The mean position of each molecule remains stationary in space but the alternate compressions and decompressions travel away

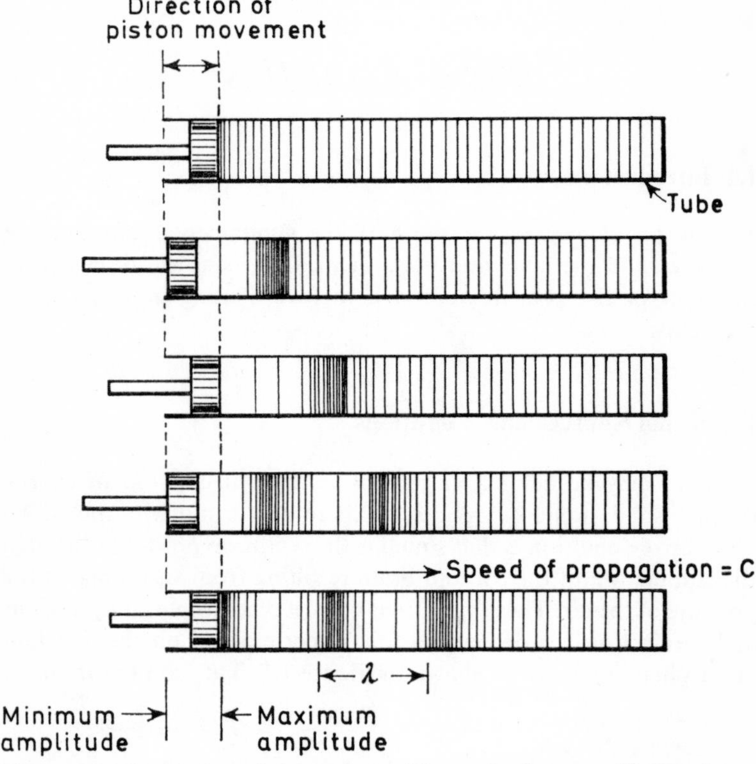

Figure 1.2. Sound waves being generated in a tube by an oscillating piston. The vertical lines divide the compressible medium in the tube into layers.

from the source in all directions if unimpeded. This process can be demonstrated with the aid of Figure 1.2 where the effect of a moving piston on the air contained in a very long tube is illustrated.

This is the simpler case of one-dimensional propagation rather than the three-dimensional case. The vertical lines represent certain layers of molecules which are equally spaced when the fluid is at rest. If the piston is pushed forward the layers of fluid in front of it are compressed. These layers will in turn compress layers farther down the tube and thus a compressional pulse travels down the tube. If, now, the piston is quickly withdrawn the layers in front of it expand and a decompressional pulse travels down the tube from layer to layer. If the piston oscillates backwards and forwards a continuous train of compressions and rarefactions or waves will be propagated along the tube (a wave consists of one compression and one rarefaction). The particles of the fluid are displaced along the direction of propagation of the wave and so it is termed a longitudinal wave. Sound waves are therefore longitudinal mechanical waves.

If a string, fixed at one end, has its free end moved up and down then a wave motion is propagated along the string, the successive parts of which vibrate in a direction at right angles to the direction of propagation. This type of vibration is referred to as transverse wave motion; surface water waves are a good example of this type of propagation.

In common with other forms of wave propagation sound waves may undergo reflection and refraction at material boundaries and can be dissipated in the form of heat within an absorptive material. Sound waves are confined to the frequency range to which the average ear responds. This frequency range extends from 20 Hz to 20 kHz and is called the audible range. (One vibration, i.e. one to and fro motion is termed a cycle and c/s have now been replaced by Hertz or Hz.) A longitudinal mechanical wave whose frequency is below the audible range is called an infrasonic wave (e.g. earthquake waves) and one whose frequency is above the audible range is called an ultrasonic wave (e.g. sonar, flow-detection).

It may be concluded that audible sound waves originate in vibrating strings (violin, guitar, vocal chords), vibrating air columns (organ, clarinet) and vibrating plates and membranes (drum, loudspeaker, xylophone). These vibrating elements alternately compress the surrounding air on the forward movement and decompress the air on a backward movement. The air transmits these disturbances outward from the source as a wave and this wave on entering the ear produces the sensation of sound. Sound cannot travel through a vacuum as there is no air to be alternately compressed and rarefied.

1.3 Velocity of Propagation

The velocity with which sound travels through the air depends on the static pressure and density of the atmosphere. At standard pressure (760 mm Mercury) and 20°C it is approximately 340 $msec^{-1}$ (1100 ft sec^{-1}). This fact is familiar through experience; during a thunderstorm the time which elapses between the lightning flash and the ensuing crash of thunder may be counted in seconds. Dividing by three is sufficient to indicate the distance from the storm in kilometres. Sonar uses a similar technique.

It has been found that sound travels much faster in other media than in air (Table 1.1).

TABLE 1.1

THE VELOCITY OF PROPAGATION OF SOUND IN VARIOUS MEDIA

Medium	*Velocity of propagation* $m\ sec^{-1}$	*Velocity of propagation* $ft\ sec^{-1}$
Air	340	1120
Water	1420	4660
Brick	3000	9900
Concrete	3300	11000
Glass	4100	13500
Aluminium	5100	16700
Steel	5200	17000

Referring back to Figure 1.2 it can be seen that some layers of molecules are in exactly the same position relative to their rest position as others. These layers of molecules are said to be in phase and the distance between two of these in-phase layers is called a wavelength λ (lambda).

There is a relationship between the wavelength λ, the velocity of propagation c and the frequency f

namely $\lambda = c \cdot \frac{1}{f}$ or $c = f\lambda$ or $f = c \cdot \frac{1}{\lambda}$

where λ = distance travelled by the wave during one complete cycle of vibration

$$f = \text{number of cycles per second (Hz)}$$

$$T = \frac{1}{f} = \text{period for a complete cycle in seconds}$$

The above relationships show that low frequencies have long wavelengths and high frequencies have short wavelengths.

The predominant component in transformer noise is the 100 Hz humming tone (twice the mains frequency) which has a wavelength of

$$\frac{340}{100} = 3{\cdot}4 \text{ metres (11 ft)}$$

The wavelength of a hissing tone of 10,000 Hz is 34 mm (approximately $1\frac{3}{8}$ in).

1.4 Octave Bands

It is convenient to divide the audible frequency range into octave bands. In each octave band the upper limiting frequency f_2 is exactly twice the lower limiting frequency f_1. The centre frequency of each octave band is defined as

$$f = \sqrt{(f_1 \times f_2)} = \sqrt{(f_1 \times 2f_1)}$$
$$= f_1 \,.\, \sqrt{2}$$

In the field of noise measurement it is necessary to be aware of the frequencies and intensities of the sounds involved. On the standard measuring instruments used the centre frequency of each octave band is given. The centre frequencies that have been internationally accepted as preferred values for measurement purposes and criteria are 31·5, 63, 125, 250, 500, 1000, 2000, 4000, 8000 Hz.

31·5 is the centre frequency of the octave band

$$\left(\frac{31{\cdot}5}{\sqrt{2}} \text{ up to } 31{\cdot}5\sqrt{2}\right) = 22{\cdot}5 \text{ up to } 45 \text{ Hz.}$$

If a finer resolution of the noise is required $\frac{1}{3}$ octave bands may be specified. The centre frequencies are related to those of the octave bands by the factor $\sqrt[3]{2}$ (cube root) and the limiting frequencies are related to the centre frequency by the factor $\sqrt[5]{2}$ (fifth root).

For example, the $\frac{1}{3}$ octave bands contained within the octave centred on 500 Hz will have centre frequencies 400, 500 and 630 Hz and the 500 Hz $\frac{1}{3}$ octave band will extend from 446 to 561 Hz.

1.5 Sound Intensity, Sound Pressure, Sound Pressure Level, The Decibel

It is known that the ear does not respond linearly to pressure. In fact, increasing the pressure fluctuation by a factor of $\sqrt{10}$ corresponds approximately to a doubling in loudness. A relationship of this type is called logarithmic, i.e. the loudness of the sound is approximately proportional to the logarithm of the pressure of the sound.

The ear which is essentially a sensitive pressure measuring instrument can just detect a pressure of 2×10^{-5} Nm^{-2} (atmospheric pressure $\simeq 10^5$ Nm^{-2}). At the other end of the range the ear can withstand sounds up to pressures of 20 Nm^{-2} approximately, at which pain begins to be sensed. This range corresponds to a pressure ratio of a million to one (10^6:1) and an intensity ratio of 10^{12}:1.

As a result of these statements it would appear to be convenient to measure the pressure of a sound on a logarithmic scale for then equal increments on this scale would result in roughly equal increments in loudness. This means that about 120 units can be used for calculations instead of a pressure ratio of 10^6:1 and it also enables addition to be used for calculations instead of multiplication.

If I_1 and I_2 are the intensities of two sources then the ratio of their intensities can be expressed as

$$N = \log_{10}\left(\frac{I_1}{I_2}\right) \text{ Bels} \qquad (1)$$

In practice it has been found more convenient to use decibels (dB) which are tenths of bels on the same logarithmic scale. Equation (1) may therefore be expressed as

$$n = 10 \log_{10} \left(\frac{I_1}{I_2}\right) \text{dB} \tag{2}$$

The decibel scale is so constructed that a difference of one "bel" between two sounds denotes a tenfold difference in intensity.

Since all modern sound detectors respond to pressure changes it is convenient to express the intensity ratio in terms of sound pressure changes. It can be shown that intensity is proportional to the square of the corresponding pressure amplitude

$$\text{i.e. } I \,\alpha\, (p^2)$$

$$\therefore \quad n = 10 \log_{10} \left(\frac{p_1^2}{p_2^2}\right) = 10 \log_{10} (p_1/p_2)^2$$

$$= 20 \log_{10} (p_1/p_2) \text{ dB}$$

Source 1 can be said to have an intensity or pressure of n decibels relative to the intensity or pressure of source 2. It will be appreciated that the decibel is a comparative unit of measurement and therefore it is most essential that a reference pressure level be specified.

For airborne sound the accepted reference level is the lowest pressure the average human ear can detect and this pressure is denoted by p_0 where $p_0 = 2 \times 10^{-5}$ Nm^{-2}.

In general the sound pressure level (S.P.L.) corresponding to the prevailing sound pressure p is given by

$$\text{S.P.L.} = 20 \log_{10} (p/p_0) \text{ dB}$$

i.e.

$$\text{S.P.L.} = 20 \log_{10} \frac{p}{0{\cdot}00002} \text{ dB}$$

Table 1.2 represents a collection of common sound pressures and their corresponding sound pressure levels.

With the aid of the S.P.L. formula and using the properties of logarithms it is possible to calculate the S.P.L. differences in decibels produced by changes in sound pressure.

It is useful to consider a few examples here.

TABLE 1.2

PRESSURE AND DECIBEL RATING OF SOME COMMON SOUNDS

Sound pressure p		*S.P.L.* (*dB*)	*Comment*	
Nm^{-2}	*μbar*			
2×10^{-5}	2×10^{-4}	0	Threshold of hearing	
		10	Soundproof room	
2×10^{-4}	2×10^{-3}	20	Ticking of a watch	Very Quiet
		30	Quiet garden	Very Quiet
2×10^{-3}	2×10^{-2}	40	Average living room	Quiet
		50	Ordinary conversation at 1 metre	Quiet
2×10^{-2}	2×10^{-1}	60	Car at 10 metres	Noisy
		70	Very busy traffic	Noisy
2×10^{-1}	2	80	Tube train, Loud radio music	Noisy
		90	Noisy factory, Heavy lorry at 5 metres	Very Noisy
2	20	100	Steel riveter at 5 metres	Very Noisy
		110	Thunder, artillery	Very Noisy
20	200	120	Threshold of feeling	INTOLERABLE
		130	Aeroplane propeller at 5 metres	INTOLERABLE
200	2000	140	Threshold of pain	INTOLERABLE
		150	White noise causing immediate deafness	INTOLERABLE
		200	Atlas rocket launch (100 m away)	

Note: Each factor of 10 increase in sound pressure results in an increase of 20 dB in S.P.L.

A $\sqrt{10}$ fold increase in sound pressure corresponds to an increase of 10 dB in S.P.L.

Example 1

What effect on the level has the doubling of sound pressure?

Let S.P.L. corresponding to sound pressure $p = L_1$
Let S.P.L. corresponding to sound pressure $2p = L_2$.

Then from S.P.L. formula

$L_1 = 20 \log_{10} p/p_0$ dB

$$L_2 = 20 \log_{10} 2p/p_0 \text{ dB} = 20 \log_{10} \frac{p}{p_0} + 20 \log_{10} 2 = L_1 + 6 \text{ dB}$$

$$[\log AB = \log A + \log B]$$

∴ Doubling the sound pressure leads to an increase of 6 dB in S.P.L.

Example 2

What is the resulting level when a sound source is added to another which is equally strong?

Total S.P.L. is obtained by taking the square root of the sum of the squared sound pressures.

$$p_{TOTAL} = \sqrt{(p_1^2 + p_2^2)}$$

Here $p_1 = p_2 = p$

$$p_{TOTAL} = \sqrt{(p^2 + p^2)} = \sqrt{2p^2} = \sqrt{2}.\, p$$

$$\therefore \quad L_T = 20 \log_{10} \frac{\sqrt{(2)}p}{p_0} = 20 \log_{10} p/p_0 + 20 \log_{10} \sqrt{2}$$

$$= L_1 + 3 \text{ dB}$$

i.e. the addition of two equal sources increases the S.P.L. by 3 dB.

Two sources each of 70 dB S.P.L. give a total S.P.L. of 73 dB and not 140 dB as is expected by the uninitiated. *Similarly* it can be deduced that three equal sources lead to an increase in S.P.L. of $10 \log_{10} 3 = 5$ dB.

Example 3

What level is produced by a source of S.P.L. $L_1 = 60$ dB and a source of $L_2 = 70$ dB together.

$$L_1 = 60 = 20 \log_{10} p_1/p_0 = 10 \log_{10} (p_1/p_0)^2 \quad [2 \log A = \log A^2]$$

$$\therefore \quad 6 = \log_{10} (p_1/p_0)^2$$

$$\therefore \quad 10^6 = (p_1/p_0)^2$$

$$L_2 = 70 = 20 \log_{10} p_2/p_0 = 10 \log_{10} (p_2/p_0)^2$$

$$\therefore \quad 10^7 = (p_2/p_0)^2$$

$\therefore$ total pressure referred to p_0, $p/p_0 = \sqrt{[(p_1/p_0)^2 + (p_2/p_0)^2]}$

$$p/p_0 = \sqrt{(10^6 + 10^7)} = \sqrt{10^6}\sqrt{(1 + 10)} = \sqrt{10^6}\sqrt{11}$$
$$= 10^3 \sqrt{11}$$

$$\begin{aligned} \therefore \quad L &= 20 \log_{10} p/p_0 = 20 \log_{10} \sqrt{10^6}\sqrt{11} \\ &= 10 \log_{10} 10^6 . 11 = 10 \log_{10} 10^6 + 10 \log_{10} 11 \\ &= 60 + 10 \log_{10} 11 = 60 + 10 \times 1{\cdot}04 = 60 + 10{\cdot}4 \\ &= 70{\cdot}4 \text{ dB}. \end{aligned}$$

Note: The first source which is considerably weaker than the second together produce a total level which is very little different from the stronger source alone.

As the difference is obviously so small the weaker source for all practical purposes may be neglected.

To simplify calculations a chart is given in Figure 1.3 which enables S.P.L.s to be added where $L_1 > L_2$.

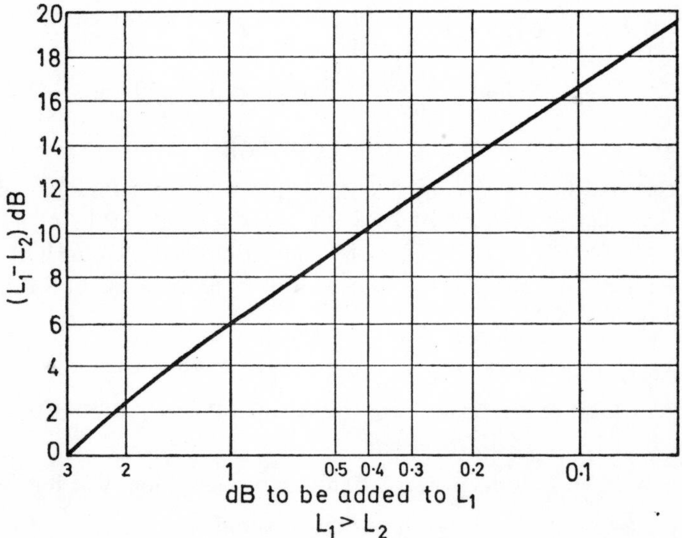

Figure 1.3. Graph for combining two acoustic levels L_1 and L_2.

In free field conditions, where the sound waves spread in a sphere of increasing radius, $I \propto 1/d^2$ and $p \propto 1/d$ where d = distance from the source. Therefore, doubling the distance from a sound source would result in quartering the intensity and halving the sound pressure which is equivalent to a 6 dB reduction in both parameters.

1.6 Level Spectrum

When making measurements on a source of noise it is frequently considered desirable to have a single reading which represents the disturbance value of the particular noise. So that this information can be obtained by a fairly unskilled operator an attempt has been made to take into account the variation in the ear's sensitivity with frequency and incorporate it electronically into an instrument. This built-in device is called the A-weighting network and measurements obtained incorporating it are referred to as dB(A). In practice all sound levels may be measured in dB(A) to give a reasonable approximation of subjective reaction to noise.

A simple instrument is sufficient to determine the level in dB(A), but the objective disadvantage is that no information about the frequency content of the sound is obtained. This is almost certainly required if any attempts are to be made to control the noise since, as will be seen later, the sound insulating properties of walls, barriers and silencers are all frequency dependent.

The extra information is obtained with the aid of a device known as a sound spectrometer which enables the S.P.L. to be measured in octave or $\frac{1}{3}$ octave bands. The level now referred to as octave or $\frac{1}{3}$ octave band sound pressure level is measured in all bands. These observations are plotted in a dB-frequency diagram as rectangular blocks as in a histogram, or more usually as points connected by straight lines as in Figure 1.4.

The narrower the frequency bands used the more information is shown by the spectrum, especially where dominant components exist in the sound. Nowadays, many instruments are available for the precise and detailed analysis of any sound. Since modern criteria for acceptable noise are available in the form of sound spectra there is no longer any need for such abstract concepts as loudness and loudness level.

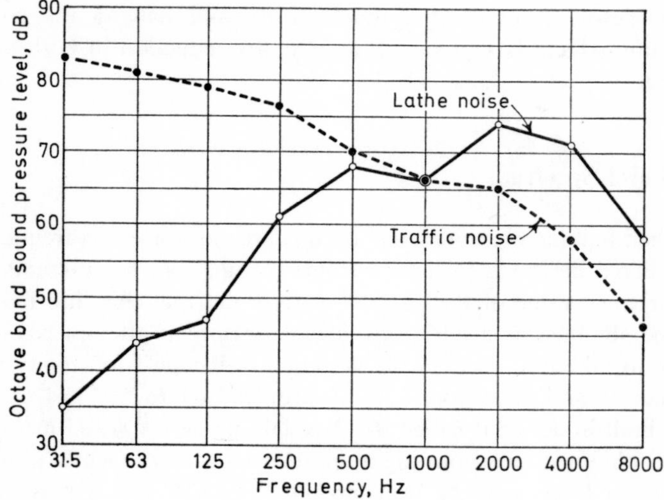

Figure 1.4. Octave band spectrum.

CHAPTER 2

Airborne Sound Insulation

2.1 Introduction

Sound generation mechanisms can be divided into two general groups. One group contains those sources which act directly on the surrounding air and rely on the air to transmit the sound. Examples of this group are the human voice, loudspeakers and musical instruments in the violin, woodwind and brass sections. The sound produced by such sources is said to be airborne sound and the insulation against such sound is logically called airborne sound insulation.

The other sound generation group comprises those sources which act directly on the structure of buildings and transmission is then through and from the structure. Examples are footsteps, noisy plumbing, furniture repositioning and banging doors. Since the majority of structure borne noises result from impacts with the structure, the group is classified as impact sound generation and insulation against it is therefore impact sound insulation.

2.2 Airborne Sound Transmission Mechanism

In order to understand sound insulation it is helpful to appreciate first the mechanism by which airborne sound is transmitted through a wall. Sound waves consist of travelling groups of rapid pressure variations and when these fluctuations act on a wall they force it to vibrate. With particularly intense sound waves it is even possible to feel the wall vibrating. The movement of the wall, however imperceptible, acts on the surrounding air in exactly the same way as a loudspeaker and sound is transmitted into adjoining rooms. The technique of airborne sound insulation, therefore, is to reduce the transmission by providing a wall which will not move easily.

2.3 Airborne Sound Insulation Units

Airborne sound insulation means reducing the energy transmitted from one point to another and therefore it is expressed as the ratio of the energy falling on a wall to that transmitted by it. Because a very large range of energy ratios is experienced in practice, it is more convenient to express the ratio in decibels and the insulation is then called the Sound Reduction Index R (Transmission Loss TL in America). The higher the decibel number the greater is the insulation and vice versa. Typical average values are given for various materials in Table 2.1.

TABLE 2.1

TYPICAL AVERAGE INSULATION VALUES

Construction	*Weight*	*Insulation dB*
460 mm (18 in) Brick plastered	950 kg m^{-2} (190 lb ft^{-2})	55
230 mm (9 in) Brick plastered	450 kg m^{-2} (90 lb ft^{-2})	50
100 mm (4 in) Dense concrete	250 kg m^{-2} (50 lb ft^{-2})	45
75 mm (3 in) Clinker block plastered	150 kg m^{-2} (30 lb ft^{-2})	40
50 mm (2 in) Solid gypsum plaster reinforced	100 kg m^{-2} (20 lb ft^{-2})	35
6 mm ($\frac{1}{4}$ in) Asbestos cement sheet	13 kg m^{-2} ($2\frac{1}{2}$ lb ft^{-2})	25
5 mm ($\frac{3}{16}$ in) Glass	13 kg m^{-2} ($2\frac{1}{2}$ lb ft^{-2})	20

2.4 The Use of Airborne Sound Insulation Curves and Average Values

Because of the variations of insulation with frequency the use of a single figure average Airborne Sound Insulation Index number is somewhat limited and cannot describe the variation of insulation with frequency. To do this it is better to have available either a curve or several figures at different frequencies and the accepted way is to quote the insulation in each of the $\frac{1}{3}$ octave bands centred on frequencies from 100 Hz to 3150 Hz inclusive. Another approach is to compare the insulation with standardised grading curves (*see* Chapter 9) and then to quote it as either Grade I or Grade II insulation. When a single figure average is required it is obtained from the arithmetic average of the insulation between 100 Hz and 3150 Hz. A quick guide can be obtained by using the insulation at the geometric mean frequency of 565 Hz as the average, but this is sometimes inaccurate.

2.5 Mass Law Insulation

When the structure of a wall is heavy and of very low stiffness, the insulation against airborne sound can be predicted from a theory

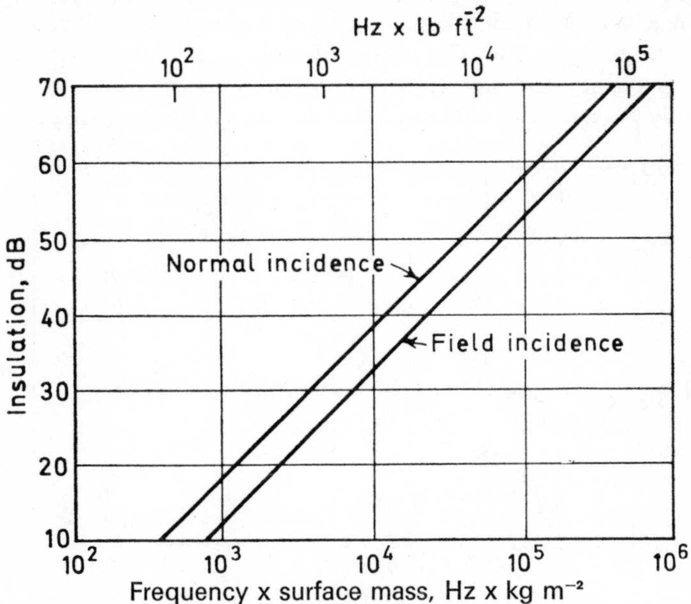

Figure 2.1. Theoretical Mass Law—insulation curves.

commonly known as the "Mass Law". This states that for each doubling of mass per unit area or for each doubling of frequency, the insulation is increased by 6 dB. The actual insulation depends on whether the incoming sound is incident at all angles or is incident only perpendicular to the wall. In practice the incoming sound is generally incident over a range of angles 0° to 80° and this is called field incidence. Curves of Mass Law insulation versus frequency and surface mass are shown in Figure 2.1, and their usage is as follows: to determine the insulation of a 50 kg m^{-2} (10 lb ft^{-2}) wall at for example 500 Hz, multiply the frequency by the surface mass. The resulting number is 25,000 (5000 or 5×10^3) and it can be seen that this intersects the field incidence line at 40 dB. Thus the insulation at 500 Hz is 40 dB and similarly is 20 dB at 50 Hz.

2.6 The Effects of Stiffness: Low-frequency Resonances

At very low frequencies stiffness cannot be ignored, indeed it is the mechanism which prevents the wall from moving when somebody leans against it. The effects of stiffness and mass both vary with frequency and, unfortunately, act in direct opposition. Consequently when the two controls are of the same magnitude they cancel

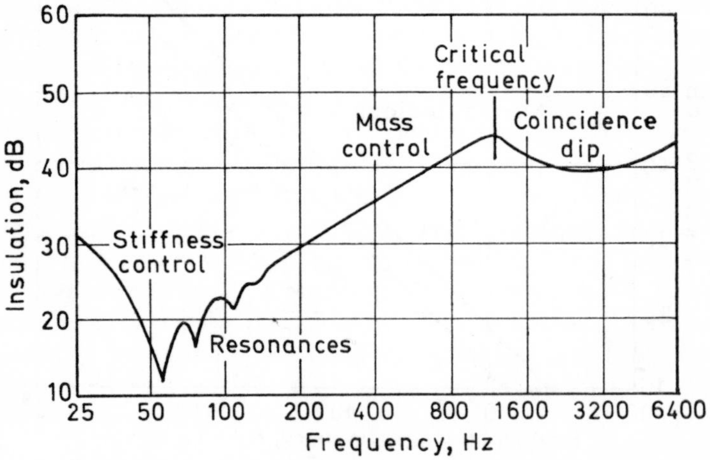

Figure 2.2. Variation of insulation with frequency.

each other out. This occurs many times because of the influence of the wall boundaries and is known as resonance. The result is one or more dips in the insulation curve at low frequencies as is illustrated in Figure 2.2.

2.7 The Coincidence Effect and its Critical Frequency

Unlike the velocity of sound in air, which is the same for all frequencies, the velocity of bending waves in a plate or wall increases with frequency. This is because bending waves are quite different in nature to compressional waves of the type transmitted in air. Bending waves are not possible in air because of the very low molecular attractions, but in solids in the form of plates or bars they transmit sound very readily.

Because the bending wave velocity varies, it follows that at some frequency it will be exactly the same as the velocity in air, for any given plate. This is known as the Critical Frequency (f_C) and generally occurs at high acoustic frequencies. When the velocities are the same it means that, since the frequency of vibration is common to both media, the wavelength in air matches exactly the wavelength of the bending wave in the plate. This exact matching gives rise to a more efficient transfer of sound energy from the air to the wall and hence a lower effective insulation for the wall. At the critical frequency the requirement is that the sound waves should be flowing along the surface of the wall (grazing incidence) and fortunately very little sound energy does this in practice so the effect is negligible. At higher frequencies when the velocity of bending waves in the wall is higher than those in air there is always some angle at which the sound wave can strike the wall such that the wavelength in the wall is matched by the effective wavelength of the sound wave as it hits the wall. An example is shown in Figure 2.3 for which the bending wave velocity is twice the air wave velocity. So at all frequencies above the critical frequency, the insulation is influenced by coincidence transmission and falls some way below the Mass Law insulation. The overall effect is shown in the complete insulation curve of a wall in Figure 2.2.

The design of a wall for high airborne sound insulation depends

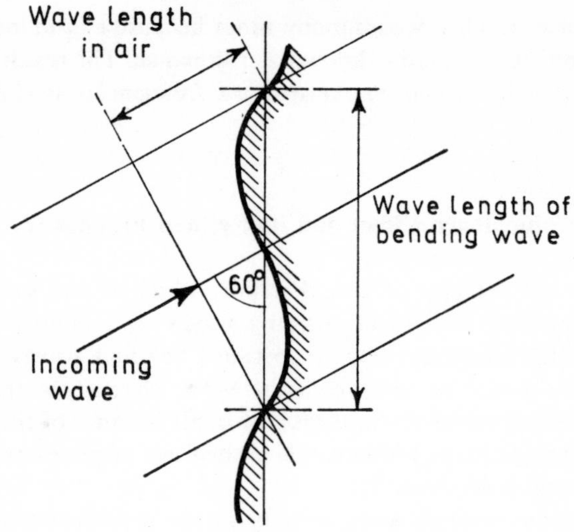

Figure 2.3. The coincidence effect.

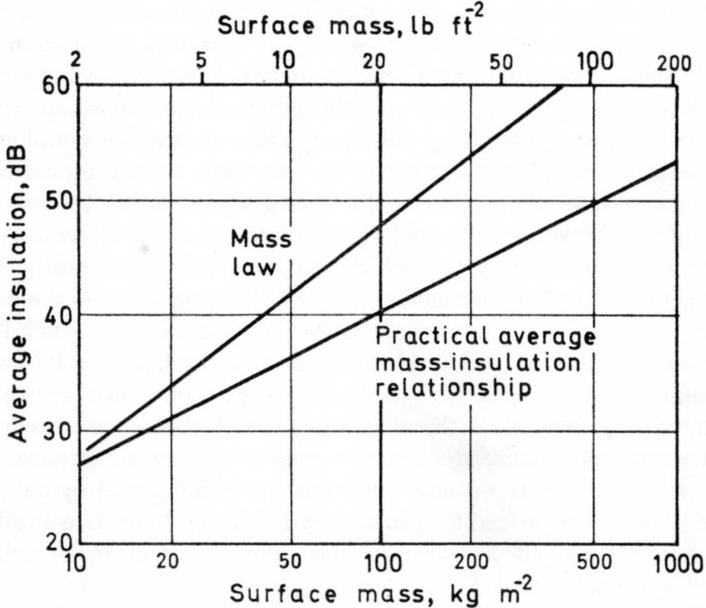

Figure 2.4. Practical average mass-insulation relationship.

therefore not only on choosing sufficient mass, but also on minimising the dynamic stiffness so that wall resonances occur at very low frequencies and also so that the critical frequency is as high as possible. These criteria, however, are the exact opposite of those generally required by the architect, therefore some compromise has to be made. Because of the stiffness effects the average insulation is generally below the mass law and the actual value can be predicted better from the empirical curves shown in Figure 2.4.

2.8 Sandwich Constructions

Partitions of sandwich construction, e.g. two sheets of hardboard with a 50 mm (2 in) straw core or two sheets of asbestos cement with a polystyrene core, have been introduced because they have a high static stiffness, are lightweight and are good thermal insulators. Acoustically they are often very poor because no thought has been given to that particular aspect.

First of all, it must be appreciated that the core is sufficiently rigid to make the construction a single leaf partition. Consequently the maximum insulation which can be achieved is determined by the mass. However, it is probable that the overall bending stiffness of the composite construction will be high, thus the insulation which is achieved in practice is quite likely to be less than the practical average mass insulation curve of Figure 2.4. To reduce the overall bending stiffness it is an advantage if the shear strength of the core is low. Then, at low frequencies, when the bending wavelength is large, the core acts as a rigid spacer so giving the construction a high stiffness. At higher frequencies, corresponding to shorter bending wavelengths, the shearing action effectively reduces the bending stiffness. This variation of stiffness with wavelength is illustrated in Figure 2.5. Also shown in this Figure is a thickness resonance, which can be troublesome if the compressional stiffness of the core is made too low. With an isotropic core material, a compromise has to be reached between low shear stiffness and high compressional stiffness and this usually means that the core thickness must not be greater than about 12 mm ($\frac{1}{2}$ in). The solution is simpler if the core is not isotropic, e.g. a paper honeycomb, because then the shear stiffness and the compressional stiffness are not directly related.

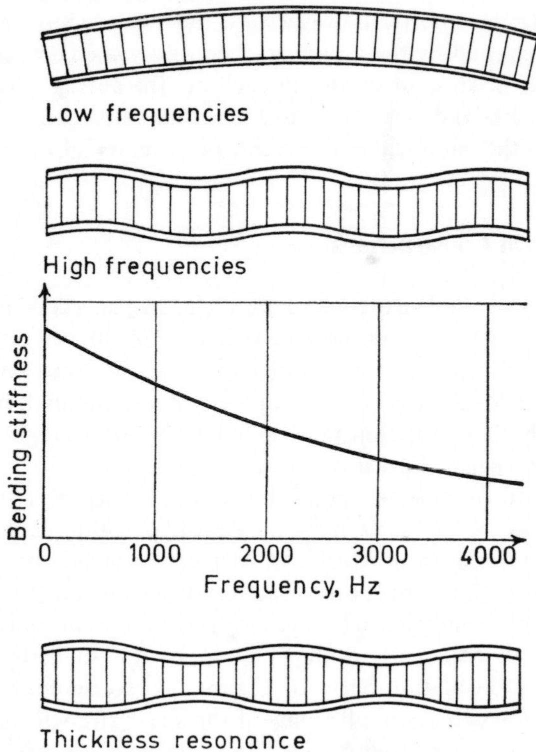

Figure 2.5. Vibrations of a sandwich construction panel.

2.9 Double Leaf Walls

Walls and partitions consisting of two leaves are quite common. They are usually referred to as cavity walls.

It might be thought that a considerable increase in sound insulation would be obtained by splitting a 230 mm (9 in) brick wall into two 115 mm ($4\frac{1}{2}$ in) brick walls separated by an air space. Each 115 mm wall has an insulation of 43 dB so a total insulation of 43 + 43 = 86 dB might be expected. In fact this is quite fallacious, because in practice the insulation would be little more than 50 dB if the walls had ties and a common footing and 60 to 70 dB if the walls were completely isolated from each other. The transmission of sound through a double-leaf structure, such as the one

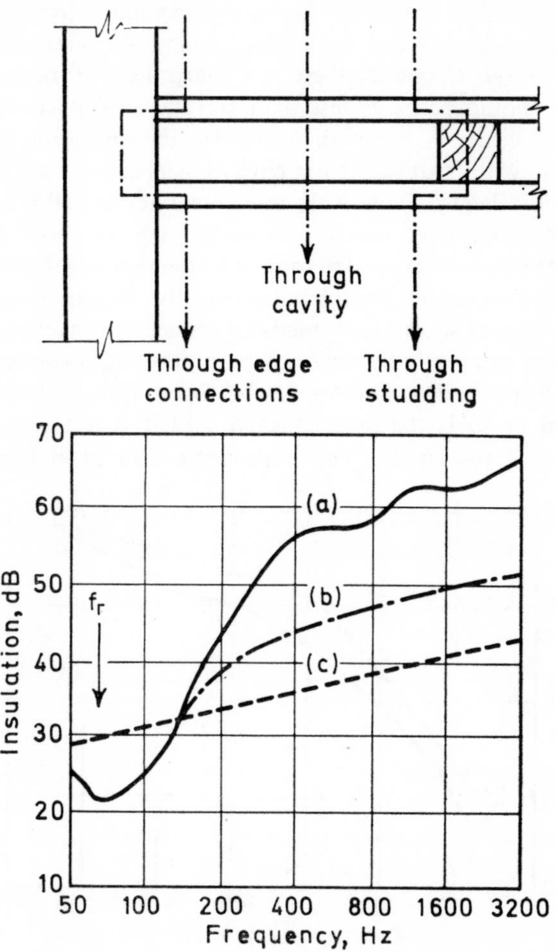

Figure 2.6. (*a*) *Expected theoretical insulation of cavity wall;* (*b*) *Effect of connections at edges and studs;* (*c*) *Equally heavy single wall.*

shown in Figure 2.6, can be through the cavity, through the studding or through the edge connections. The graph of Figure 2.6 shows the insulation that is obtained when the two leaves are entirely unconnected and also the effect of connecting studs and a common perimeter.

At low frequencies the air between the two leaves couples them together rather like a spring. There is a resonant frequency, f_R, determined by the mass of the leaves and by the thickness of the cavity and at this frequency there is a sharp drop in insulation. In practice f_R should be made less than 100 Hz, so for leaves of low mass a wide cavity is essential. Above f_R, the insulation increases more rapidly with frequency for a cavity construction than it would for an equally heavy single wall, but above about 250 Hz there is the further problem of cavity resonances. Up to about 1000 Hz the resonant frequencies are determined by the two larger dimensions but at higher frequencies the thickness may also be important. Many resonances can exist and each tends to reduce the insulation. Fortunately they can easily be removed by putting an absorbent infill, e.g. mineral wool, glass fibre or flexible foam, into the cavity. The first 25 mm or so is the most effective and it is not necessary to completely fill the cavity. The improvement in insulation which

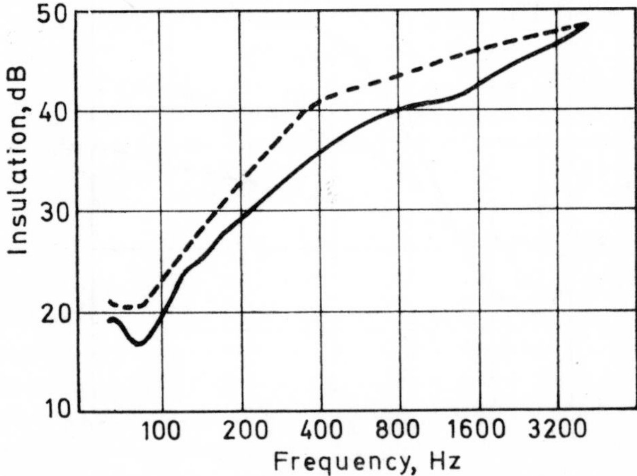

Figure 2.7. The influence of absorbent in the cavity of a double leaf partition (a) without absorbent ——, (b) with absorbent - - - - - -.

might be expected is shown in Figure 2.7, although if the individual leaves are very stiff and coincidence transmission starts to occur at a low frequency, the improvement due to the absorbent can be even more startling as is shown in Figure 2.8. Generally, an absorbent infill is most effective on lightweight constructions and has almost no effect on a cavity brick wall.

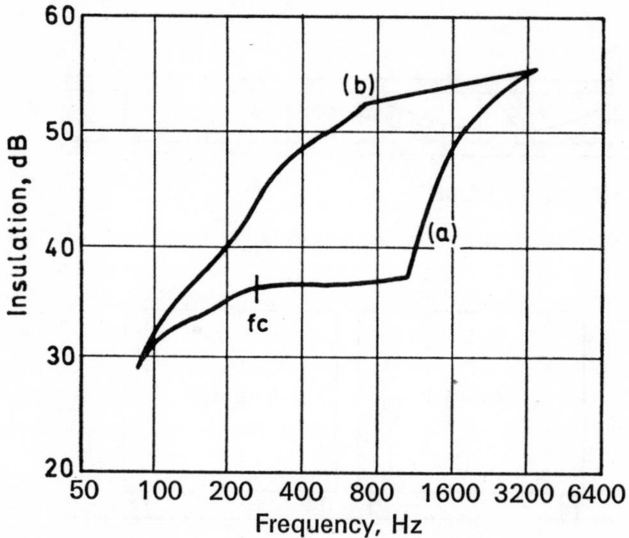

Figure 2.8. The influence of absorbent in the cavity of a double leaf partition when the critical frequency is low (a) without absorbent, (b) with absorbent.

Transmission through studding and round the common perimeter becomes important when an insulation of more than 40 dB is required. It usually affects the insulation more at high frequencies than at low frequencies. Up to 40 dB, common studding at, say, 0·6 m (24 in) pitch is acceptable, but above 40 dB staggered studding should be used so that there is no direct link between the two leaves. For an insulation in excess of 50 dB the perimeter linking becomes important and should be avoided by vibration isolating the rooms. These situations are illustrated in Figure 2.9.

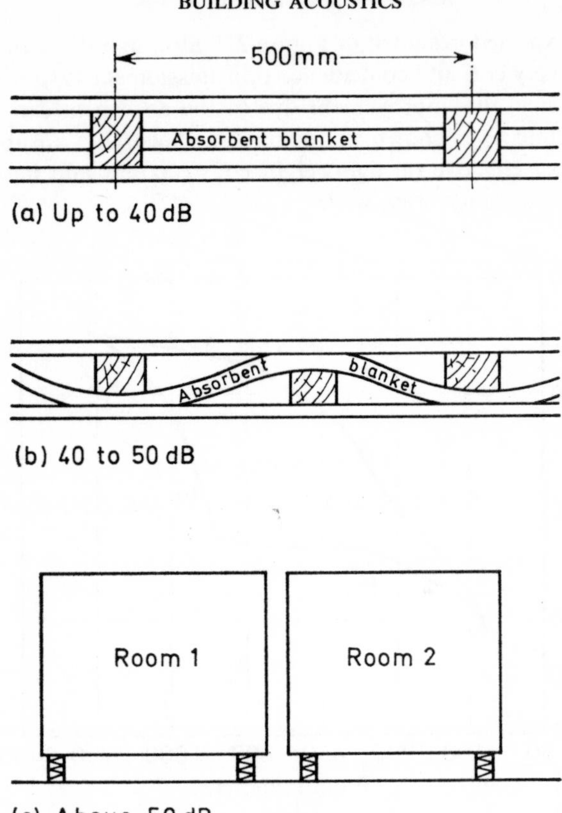

Figure 2.9. Studding in cavity walls.

Theoretical attempts to predict the insulation of a double-leaf wall have, so far, proved unsatisfactory. The most helpful guide is purely empirical and is shown in Figure 2.10. It is assumed that an absorbent infill is used and that linkage between the leaves conforms with the requirements laid down in the previous paragraph. A combination of mass and cavity thickness which gives a nominal insulation of less than 40 dB should not be used because the resonant frequency, f_R, would not be below 100 Hz and so the low frequency insulation would be very poor. If an insulation of only, say, 35 dB is required then the mass and cavity thickness giving 40 dB in Figure 2.10

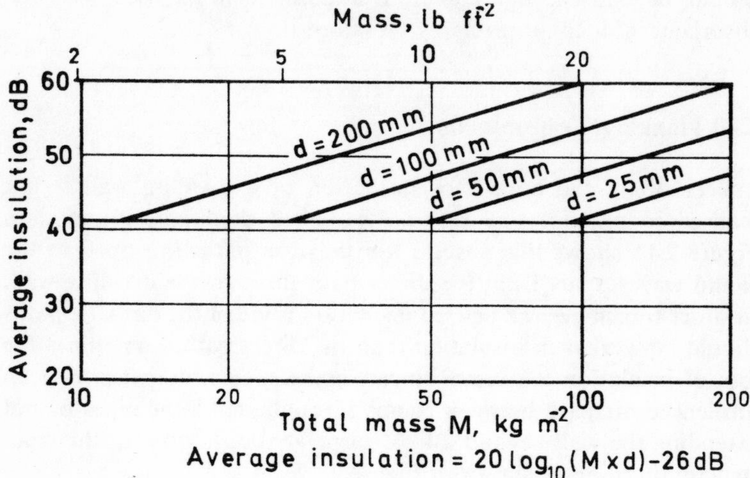

Figure 2.10. Average insulation of a cavity wall.

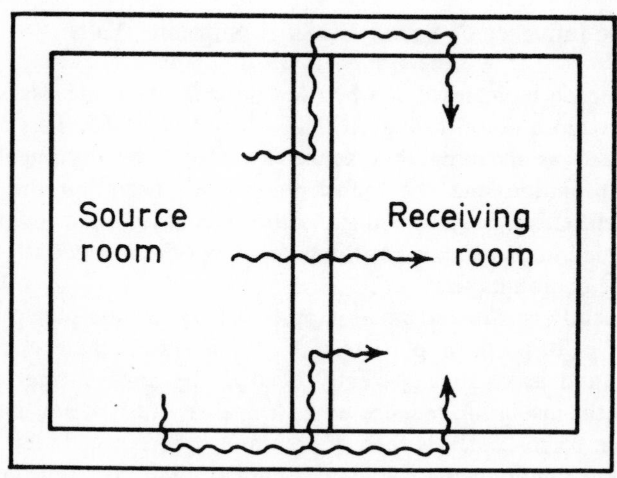

Figure 2.11. Transmission paths.

should be chosen, but common studding can be used and the absorbent infill could probably be skimped.

2.10 Flanking Transmission

Very often, the hoped for insulation of a dividing wall is not realised in practice due to what is called flanking transmission. Figure 2.11 shows that several transmission paths are open to the sound waves apart from the direct path through the dividing wall. In order to achieve the desired insulation none of the flanking paths should be weaker in insulation than the direct path. Sometimes the loss of insulation is caused by carelessness, such as permitting an unsilenced air path between rooms through ventilator pipes or not extending the wall beyond a light suspended ceiling up to the main structural ceiling as shown in Figure 2.12.

The consideration of flanking paths also means that in some cases the addition of a high quality insulating wall would be pointless without making major structural changes in other areas such as ceiling or floors.

2.11 The Influence of Sound Leaks; Composite Walls

Experience indicates that when the door is open one hears considerably more noise through it than when it is closed. This can be interpreted as meaning that sound leaks may be detrimental to sound insulation and the higher the sound insulation the more serious the effect. In spite of this the above facts are often completely disregarded in the case of partitions dividing offices, especially when these are demountable.

Frequently cracks and small gaps exist between component parts of this type of partition particularly at floor and ceiling level. Sometimes sound leaks are apparently "sealed" by an ill-fitting tongue but still the insulation remains poor. A decrease of 10 dB in average, e.g. from 35 to 25 dB, is then of common occurrence. Later, when discussing requirements for good sound insulation it will be seen that the above-mentioned situation can prove disastrous.

In general a wall is composed of parts which have widely differing insulations; a wall with a door, etc. The insulation can be determined theoretically but is more easily interpolated from Figure 2.13. Frequently the insulation of the entire wall has a much lower insulation than the component possessing the highest insulation.

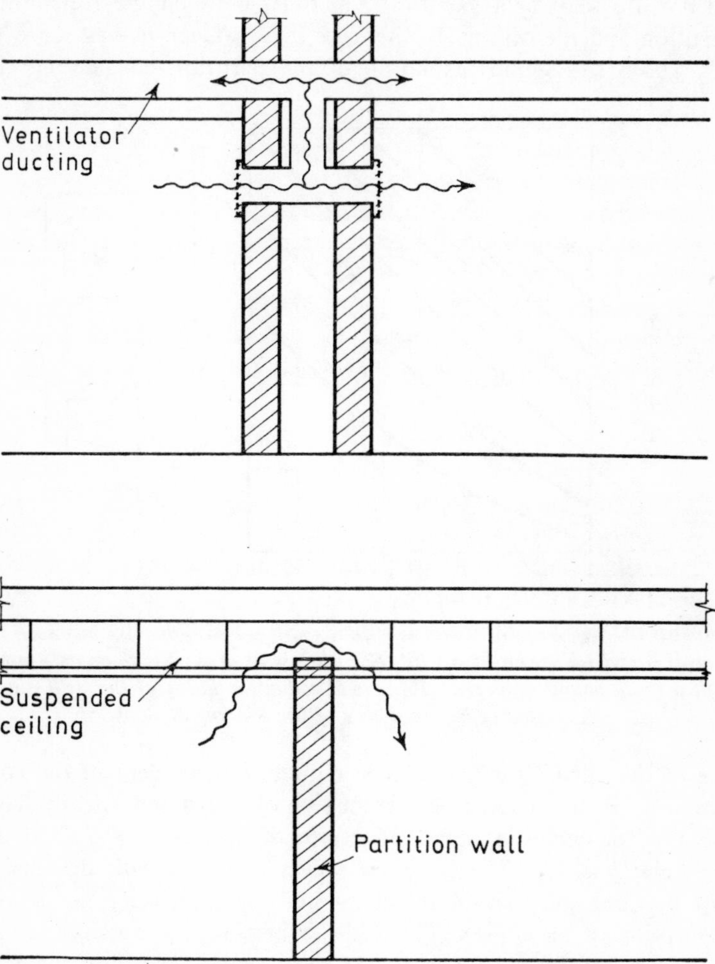

Figure 2.12. Two examples of insulation errors.

Therefore it is pointless to introduce a normal type of domestic door into a wall with high sound insulation. The weakest link determines the strength of the whole chain.

A peculiar type of composite wall is the *partial partition*. These are partitions 2·1 m (7 ft) high used to subdivide large office spaces into smaller accommodation areas. The ceiling height is 3·4 m (11 ft) thus leaving a gap of 1·3 m (4 ft) between the top of the partition and the ceiling. In this case it no longer makes sense to talk about the sound insulation of the partition considering the

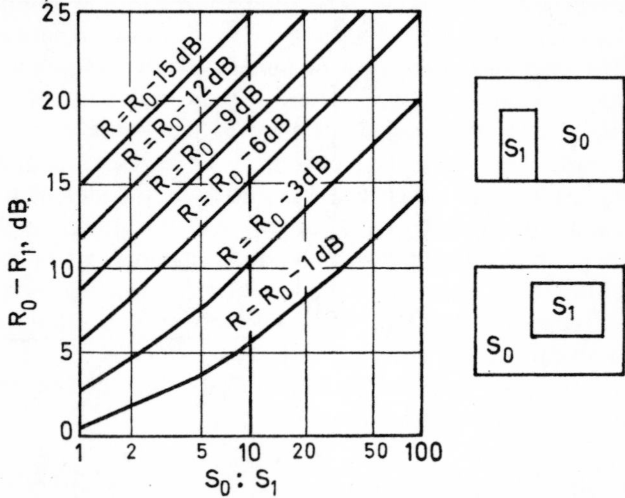

Figure 2.13. Influence of surfaces with a small sound reduction index on the overall sound reduction of a wall. S_0: total wall area; S_1: door or window with a small sound reduction, R_0: sound reduction index of the wall alone; R_1: sound reduction index of the window or door.

size of the "leak" for it is almost entirely independent of the construction. Even so some benefit may be obtained acoustically from this type of partition, since reductions of the order of 5 to 10 dB are quite possible. The beneficial effect increases with decreasing gap between partition and ceiling and by increasing the sound absorption of the ceiling. The latter determines the possible reduction in sound level and so a first-class sound-absorbing material should be used.

2.12 Doors

It is a mistake to assume that, because a door is constructed from a given material, its sound insulation will follow the experimental Mass Law for that material. In general, it falls some way below it. This is due to the fact that even if the door were perfectly sealed round all four edges its size is very much smaller than the average partition, so its low frequency resonances occur at much higher frequencies than in a wall made from the same material. However, the other basic principles which have already been discussed in connection with the sound insulation of single and double leaf partitions still apply in that a door constructed, say, from two thin sheets of steel separated by a 50 mm (2 in) air gap containing a mineral wool quilt will have a much higher insulation than a door constructed as a single leaf of steel but of equal weight.

The importance of sealing cannot be overstressed and this invariably leads to the necessity for a threshold, a component which has found little popularity in Britain. On the other hand, it is possible to obtain fairly good insulation in situations where the door must swing truly. This can be achieved by constructing both frame and the edges of the door from steel channeling, filling with a porous absorbent and covering with a perforated sheet metal. It is essential that the door fits accurately into the frame so that the clearance always remains the same.

Examples of Doors

45 dB	Two 50 mm (2 in) solid wood doors with all cracks adequately sealed in conjunction with a sound lock.
35 dB	Two flush panel doors (hollow core with 3 mm ($\frac{1}{8}$ in) hardboard or plywood both sides) with cracks adequately sealed and in conjunction with a sound lock.
25 dB	50 mm (2 in) solid door with normal cracks around edges.
15 dB	Flush panel door with normal cracks around the edges.

2.13 Outer Walls and Windows

All too frequently today people are disturbed by noise reaching them from outdoors, caused by air and road traffic and by industrial activities. It is therefore appropriate to discuss the sound insulation of outer walls and roofs. The latter present a lesser problem since many modern buildings have concrete roofs or roofs made from comparably heavy materials. Only when a timber construction is used and when windows or so-called sky domes are introduced is the insulation of the roof less than that of the outer walls containing windows.

The insulation of an outer wall is usually but not necessarily determined by the insulation of the windows. In some types of curtain walling the unglazed panels are so light in weight, being made for example from two sheets of asbestos bonded together, or wood fibre board, that their insulation is even inferior to that of the glazed area. In other types of outer wall it is the windows which form the weakest link in the insulation. If the windows are made to open then the sound leaks so created largely determine the sound insulation, irrespective of the type of glazing. For a high degree of insulation it is essential that fixed windows are used with adequate sealing of any cracks between the frame and masonry. Of course, when fixed glazing is used then mechanical ventilation becomes a necessity.

The sound insulation provided by a window follows the same laws which have been mentioned above. Mass per unit area and bending stiffness play their part just as in the case of a wall, but in addition the angle of incidence of the sound waves has to be considered. In the case of the wall dividing two rooms it is taken for granted that sound waves strike the wall at completely random angles of incidence. In the instance of sound from outdoors there is generally a predominating angle of incidence. For this reason data on the insulation of windows is given for different angles, e.g. 0° (normal incidence) 45° and 70°. The latter angle gives the lowest insulation value (*see* Figure 2.14).

The influence of the coincidence effect can be seen in all of the curves and naturally the frequency at which coincidence transmission begins is lower for a thick pane than a thin one. Coincidence starts at around 600 Hz for a very thick pane. Note the critical frequency cannot be found as this is the lowest frequency at which coincidence occurs for grazing incidence, i.e. at an angle of incidence of 90°.

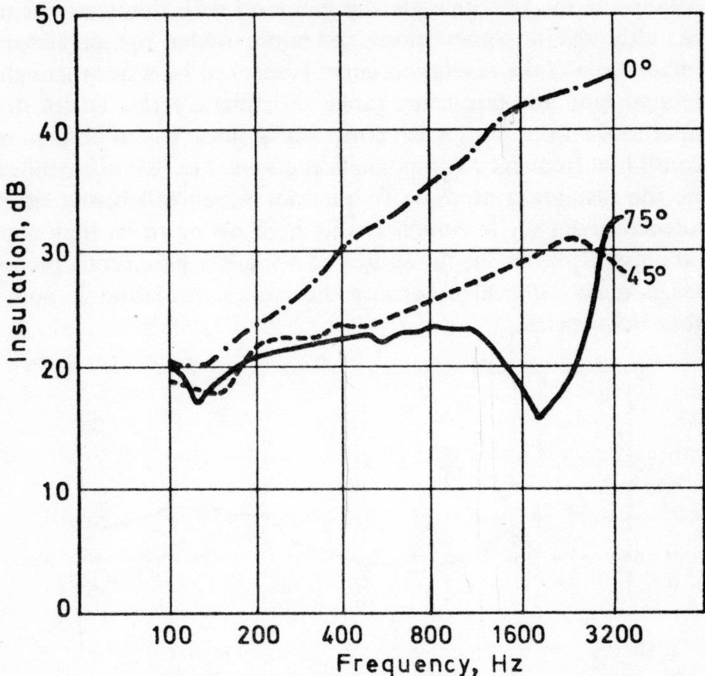

Figure 2.14. The effect of the angle of incidence on the sound insulation of single glazing.

At 75° or 70° the effect is present from a higher frequency onwards. Examples are shown in Table 2.2.

TABLE 2.2

Thickness of pane mm	*Approximate critical frequency Hz*	*Angle of incidence degrees*
12	1000	90
	1040	75
	1420	45
8	1500	90
	1560	75
	2130	45

The insulation of single glazing increases with the mass per unit area, although it comes more and more under the influence of coincidence, i.e. the insulation curve is marked by a deep trough in the important mid-frequency range. Fortunately this is not detrimental to the insulation when street traffic noise has to be kept out, because low frequency components predominate. Even for a heavy pane the insulation at those frequencies which fall below the coincidence frequency is comparatively high owing to its high superficial mass. If, however, the source of noise is a jet aircraft then the noise is more noticeable because the pane's insulation is poor at higher frequencies.

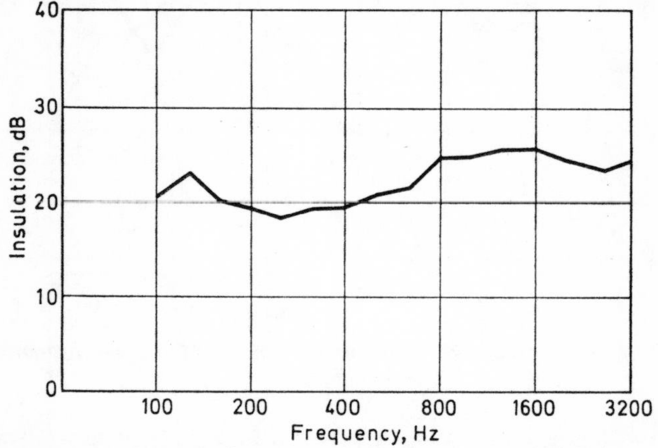

Figure 2.15. Double glazed sealed unit with very small air gap.

Two panes of glass separated by a layer of air constitute a double-leaf construction. Its behaviour is explained by the mass-spring mass analogy, characterised by the resonant frequency, f_R. The resonant frequency depends on both the total mass and the stiffness of the spring. Since in double glazing the masses are relatively small, the stiffness of the spring should be low, i.e. the thickness of the air layer should be large, in order to make the resonant frequency sufficiently low. The double glazing which is used for purposes of thermal insulation, of the sealed unit type, has a very narrow cavity of between 10 and 12 mm, so that the masses are coupled through a stiff spring. As a result, the resonant frequency lies at approximately

300 Hz and consequently the insulation in this frequency region is disappointingly small (*see* Figure 2.15).

This is a serious disadvantage in this type of double glazing if it is intended as an insulator against traffic noise as well as a thermal insulator. Compared to a single airtight pane of glass 6 mm thick a sealed double glazed unit with 6-mm panes and a 10-mm air space is actually inferior.

It is now obvious how double glazing can be made efficient as a

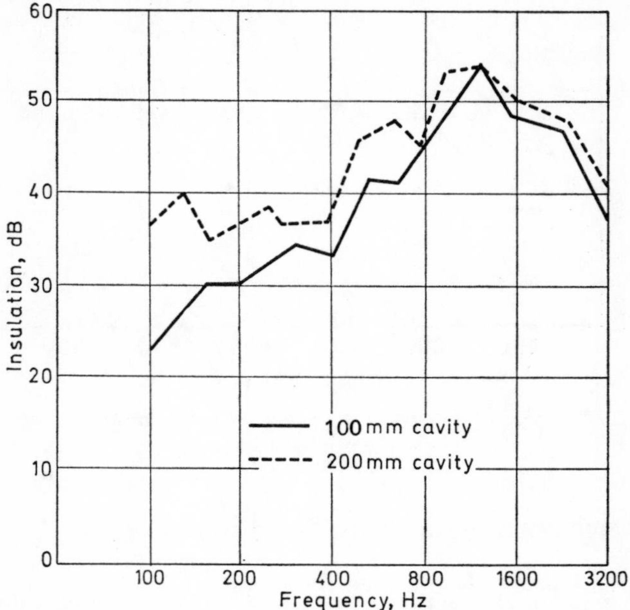

Figure 2.16. Influence of the cavity size on the insulation of double glazing.

sound insulator: that is by increasing the width of the cavity to 75 mm (3 in) or 100 mm (4 in) or even more if possible. In Figure 2.16 two examples are shown.

A further improvement is obtained by attenuating the air resonances which can occur between the sides of the double glazed window frame by placing some sound absorbing material on them. The influence of the absorbent placed round the reveals can be seen by comparing the two curves in Figure 2.17.

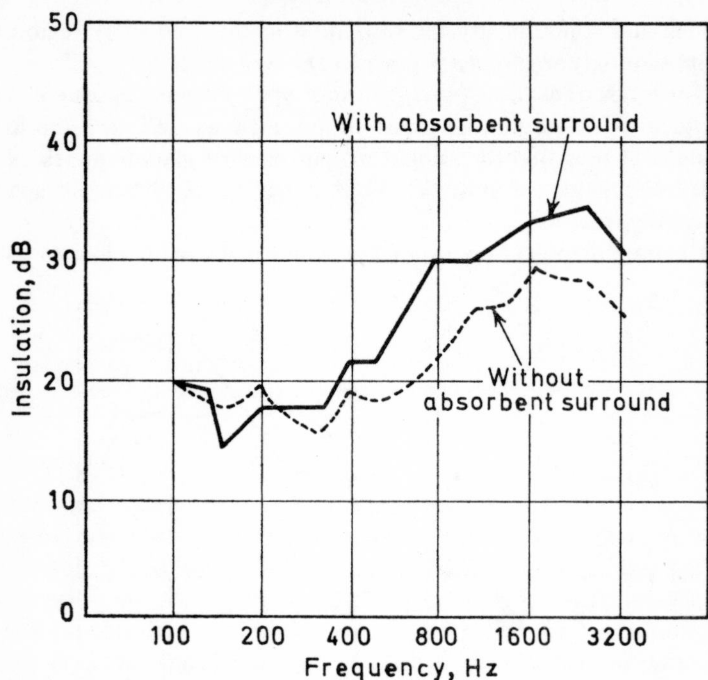

Figure 2.17. Effect of absorbent surround on sound insulation of double glazed window.

2.14 Requirements for Airborne Sound Insulation

2.14.1 *Walls and Floors between Dwellings*

Table 2.3 shows the subjective impressions which arise with dividing walls of different insulation values.

It also gives a rough indication of the necessary requirements in different circumstances. It should be noted that the table goes in steps of 5 dB, a figure which can be taken to indicate an improvement (or lowering) of the classification of the particular wall.

The requirements of airborne sound insulation of walls and floors in dwellings has received most attention. These are specified in many countries in the form of *Codes of Practice* and *Grading Curves*.

TABLE 2.3

Average insulation in dB 100 *to* 3150 *Hz*	*Subjective assessment for speech and music in a neighbouring room*
60	Loud radio inaudible
55	Radio at normal volume inaudible
50	Radio at normal volume just audible
45	Loud speech just intelligible, melodies can be recognized
40	Conversation at normal level just intelligible
35	Normal conversation clearly intelligible
30	Sounds as though a radio at normal volume is playing softly in one's own room

The British Standard Code of Practice, CP3 Chapter III (1960) gives the insulation grades for dwellings. The grades are in the form of curves (*see* Chapter 9). The house party-wall grade is based on the performance of a 230 mm (9 in) brick wall. The degree of insulation provided by this wall appears to reduce the noise from neighbours to a level that is acceptable to the majority. Grade I refers to an insulation where the noise from neighbours only gives rise to a minor disturbance. It is considered to be no more of a nuisance than many of the other disadvantages which tenants may associate with living in flats. With Grade II insulation the neighbours' noise is considered to be the worst feature of living in flats but even so at least half the tenants are not seriously disturbed. If the insulation is worse by 8 dB or more than Grade II then the noise from neighbours is often considered intolerable and is likely to lead to serious complaints.

Table 2.4 gives a few examples of walls and constructions, their average sound insulation from 100 to 3150 Hz and their grading.

2.14.2 *Partitions in Schools and Offices*

Partitions separating classrooms should have an average insulation of approximately 40 dB provided that the classroom does not have an excessive reverberation time, i.e. has a maximum reverberation time of 1 sec. This means that there is no necessity for a heavy

230 mm (9 in) brick wall, a 115 mm (4½ in) brick wall would be quite adequate. Doors in the wall are to be avoided and folding or sliding partitions are in general acoustically unsatisfactory.

TABLE 2.4

Construction	*Average sound insulation (dB)* 100 *to* 3150 *Hz*	*House party-wall grade*
230 mm (9 in) brick with 12 mm (½ in) plaster both sides	50	I
180 mm (7 in) dense concrete with 12 mm (½ in) plaster both sides	52	I
Double 50 mm (2 in) clinker block with 150 mm (6 in) cavity. No ties. 12 mm (½ in) plaster both sides	50	I
Double 100 mm (4 in) clinker block with 50 mm (2 in) cavity. Thin wire ties. 12 mm (½ in) plaster both sides	50	I
Double wall with 230 mm (9 in) brick leaf and 115 mm (4½ in) brick leaf. 50 mm (2 in) cavity. Thin wire ties. 12 mm (½ in) plaster both sides	53	I
205 mm (8 in) hollow dense concrete block with 12 mm (½ in) plaster both sides	45	II
Double 50 mm (2 in) clinker block with 50 mm (2 in) cavity. Thin wire ties. 12 mm (½ in) plaster both sides	47	II

Partitions in offices should guarantee privacy. It has been shown that this is not only dependent on the sound insulation of the partition but also on the level of background noise coming from the street due to traffic and noise from the air conditioning system if one exists. It is therefore impossible to give exact figures for the required minimum insulation since this is determined by the actual situation.

Owing to a sound leak which is sometimes unavoidable the popular moveable partition may be very unsatisfactory. In many cases such partitions have been installed with average insulations of between

20 dB and 25 dB giving rise to many complaints about lack of privacy, the overhearing of telephone conversations, etc.

In a normal quiet office where rooms are used for confidential discussions an average insulation of at least 40 dB is advisable. In fact, the average value does not give a fully reliable indication of the ability of the partition to safeguard privacy. The insulation as measured in octave bands centred at 500, 1000 and 2000 Hz is much more important than at lower and higher frequencies.

2.15 Design for Maximum Sound Insulation

For single leaf walls the maximum insulation which can be achieved is given by the Mass Law (*see* Figure 2.4). Normally coincidence transmission occurs below 3000 Hz so that the average insulation falls somewhat below the theoretical maximum. The situation is worse with heavy walls than with light walls because heaviness involves a greater thickness and therefore a greater stiffness leading to a lower critical frequency. Clearly, if the object is to avoid coincidence transmission, some materials are more suitable than others. Table 2.5 lists some materials in order of preference.

The first column gives the critical frequency multiplied by the surface density and the second gives the surface density per mm thickness. So if it is decided that f_C should not be less than 3000 Hz, the maximum permissible surface density of, for example, asbestos cement is 11·2 kg m^{-2} (2·3 lb ft^{-2}) corresponding to a thickness of 5·8 mm (0·23 in). This would give an average insulation of 28 dB. A thicker sheet, say 11·6 mm (0·46 in) thick, would give a greater insulation but because of coincidence, now occurring at 1500 Hz, the extra weight would not be working at maximum efficiency and the average insulation would only be about 31 dB. It would be much better to use two sheets, each 5·8 mm thick, adjacent but not glued together. Then coincidence would still not occur below 3000 Hz and the insulation would be 34 dB.

For a composite partition of two materials or even three when the core material is stiff, e.g. flaxboard, the problem may be approached more fundamentally. The Young's modulus of elasticity and the density may either be found from tables or may be measured. Using these, the composite bending stiffness and hence the critical frequency may be calculated. Even without such calculation it is obvious from Table 2.5 that some combinations would not be suitable. Flaxboard, for example, should not be used on its own with a thickness of more than 25 mm (1 in), so it would clearly not be wise to face it with, say, hardboard or plywood. On the other hand, a facing of lead or partition board might be acceptable because both of these would add more mass than stiffness and so would tend to increase the critical frequency of the flaxboard.

TABLE 2.5
CRITICAL FREQUENCIES OF SOME COMMON MATERIALS

Material	*Critical frequency × surface density Hz × kg m⁻² (Hz × lb ft⁻²)*	*Surface density per unit thickness kg m⁻² mm⁻¹ (lb ft⁻² in⁻¹)*	*Young's modulus Nm⁻² (lb in⁻²)*
Lead	600,000 (123,000)	11·2 (58)	14×10^9 (2,100,000)
Partition board	124,000 (25,500)	1·6 (8·5)	$1{\cdot}1 \times 10^9$ (160,000)
Steel	97,700 (20,000)	8·1 (42)	210×10^9 (30,000,000)
Reinforced concrete	44,000 (9,000)	2·3 (12)	24×10^9 (3,500,000)
Brick	42,000 (8,400)	1·9 (10)	16×10^9 (2,300,000)
Glass	38,000 (7,800)	2·5 (13)	41×10^9 (6,000,000)
Asbestos cement	33,600 (6,900)	1·9 (10)	24×10^9 (3,500,000)
Aluminium	32,200 (6,600)	2·7 (14)	69×10^9 (10,000,000)
Hardboard	30,600 (6,300)	0·81 (4·2)	$2{\cdot}1 \times 10^9$ (300,000)
Plasterboard	29,200 (6,000)	0·75 (3·9)	$1{\cdot}9 \times 10^9$ (270,000)
Plywood	13,200 (2,700)	0·58 (3·0)	$4{\cdot}3 \times 10^9$ (620,000)
Flaxboard	13,200 (2,700)	0·39 (2·0)	$1{\cdot}2 \times 10^9$ (170,000)

With a three-layer composite partition containing a relatively soft core, e.g. expanded polystyrene, the shear stiffness of the core also plays an important part and the calculation, although possible, is much more difficult. It is probably easier, and perhaps more certain, to make up a small sample of length L and measure the

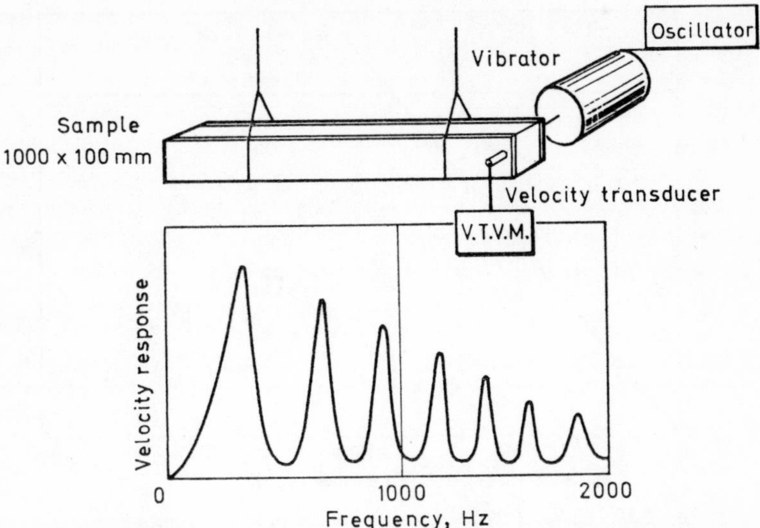

Figure 2.18. Method of measuring velocity of bending waves.

bending wave velocity as a function of frequency. This is done by suspending the sample from strings, as is illustrated in Figure 2.18, shaking it at one end with a vibrator and monitoring the response with an accelerometer. The fundamental resonance can be excited very easily and the next ten or more resonances can usually be observed. The frequencies of these resonances, f_n, are noted and the bending wave velocity at each frequency is obtained from the equation

$$C_B = \frac{Lf_n}{n/2 + 1/4} \quad (n = 1, 2, 3, \text{etc.})$$

When C_B^2 is plotted as a function of frequency it is seen to increase (Figure 2.19) and the critical frequency, f_C, is given by the point at which C_B is equal to the velocity of sound waves in air. If necessary the line will have to be extrapolated from the maximum measurement frequency. If the shearing action of the core plays no part, or a single or double layer partition is being tested, the points should all fall on a straight line (Figure 2.19) but with a soft core, the shearing action will cause the line to curve away to the right (Figure 2.20).

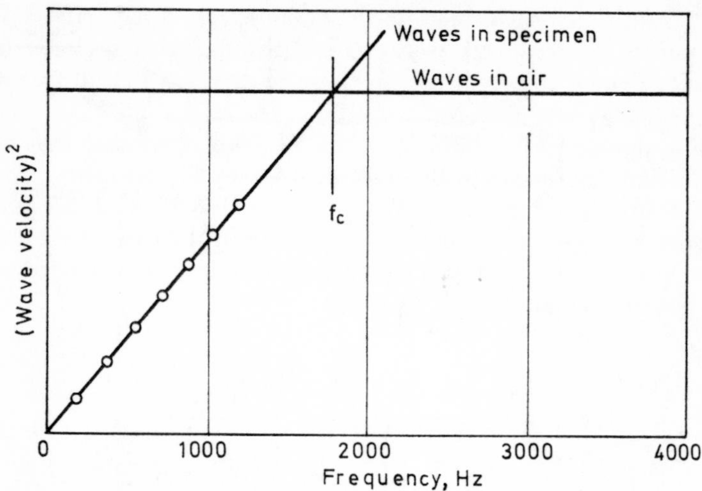

Figure 2.19. Critical frequency of a specimen with no shearing action.

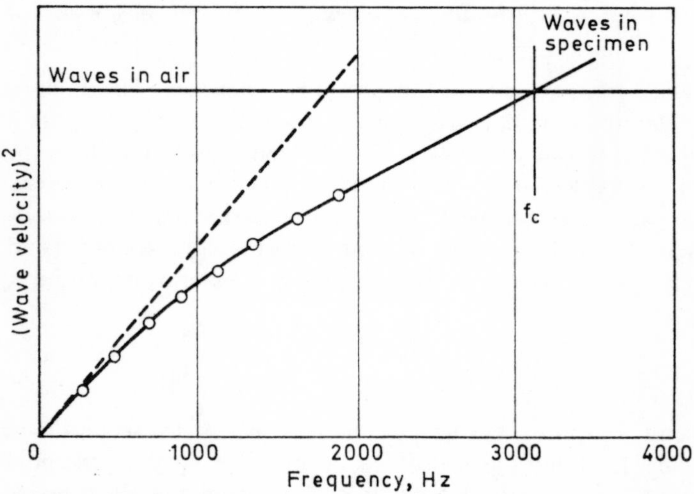

Figure 2.20. Critical frequency of a specimen with shearing action.

If it is thought that the thickness resonance of a three-layer partition might be a problem, because the core is relatively soft, a small square sample should be mounted on a vibrator as is illustrated in Figure 2.21. An accelerometer mounted on top monitors the output as the frequency is varied and the thickness resonant frequency can be observed. It is sufficient to establish that it does not occur below, say, 3000 Hz.

For double-leaf partitions, it can in general be said that any insulation deficiencies in the individual leaves will also show up in the double-leaf insulation. With two good individual leaves, it should be possible to achieve the insulation indicated by Figure 2.10, provided that flanking transmission and leaks are avoided and an absorbent infill is used.

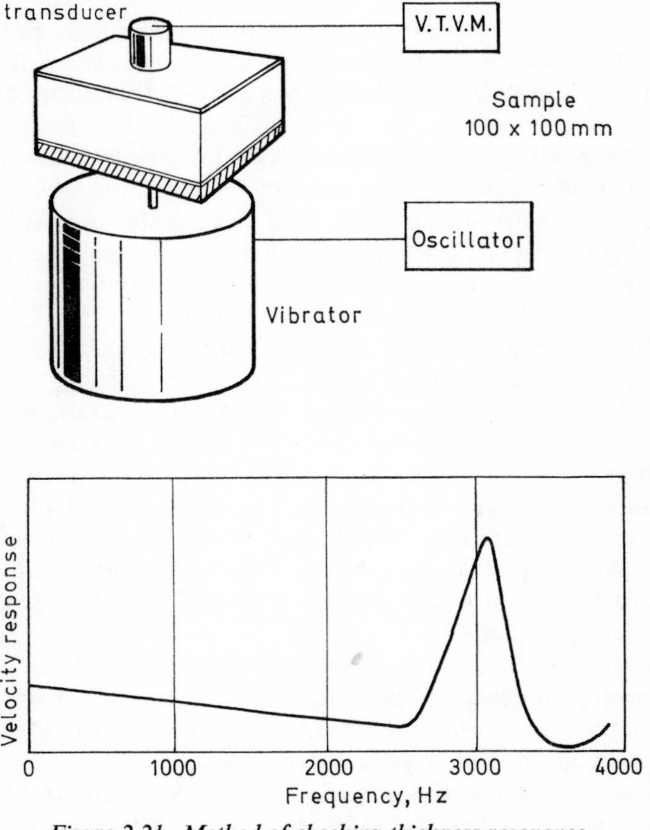

Figure 2.21. Method of checking thickness resonance.

CHAPTER 3

Impact Sound Insulation

3.1 Introduction

If sound from a source is transmitted to a listener through the air only, it is termed airborne sound. Structure-borne sound is the term used for sound which is produced when energy is transmitted directly to a structure, e.g. by a piece of machinery, and this structure then radiates the energy as sound. In addition, if the energy is transmitted intermittently to the structure, e.g. by footsteps, the resulting sound is termed impact sound. With both structure-borne and impact sound the energy must be radiated into the air by perhaps a wall or ceiling before it can be heard. Structure-borne sound and impact sound are identical except in their production.

With impact noise it is possible to obtain large displacements of the structure and when the internal damping is low (e.g. concrete) these displacements are propagated through the structure with little attenuation. Once the energy is in the structure, large areas can be set into vibration giving rise to a high degree of radiated noise. Sometimes the large sound radiating area is useful, e.g. with musical instruments such as cello, double bass, etc. However, in many cases this phenomenon is undesirable—a vibrating heating pipe or water pipe firmly fixed to a wall or floor will radiate far more noise than if it were vibrating alone.

The means of insulating structure-borne noises are different from those of suppressing airborne noises. It is therefore important to find out whether the noise that has to be reduced originates from airborne sounds or from structure-borne sounds.

Since impact noises are readily transmitted through building structures every effort should be made to prevent energy from the source from reaching the main structure. This may be accomplished by (*a*) the use of adequately resilient flooring (carpeting, rubber tiles, cork tiles, etc.) to reduce impact transmissions to the floor, and (*b*) the use of flexible mountings, anti-vibration pads, floating floors, etc.

3.2 Structural Damping and Discontinuities

In the case of floor constructions the internal damping is generally low and the bending waves are freely transmitted through the whole structure with little attenuation. The use of damping mechanisms could theoretically have a useful effect, but when the problem is considered it is found to be impracticable for structural or economic reasons. For example, in a 150 mm (6 in) concrete floor the amount of damping material required would be a layer at least 100 mm (4 in) thick to have any marked effect. In the case of non-floor constructions such as steel plates damping materials may be used beneficially. Discontinuities of structure provide some reflection of the bending wave and hence an attenuation of the transmitted wave, but the attenuation is quite small, generally not exceeding 6 dB. Discontinuities may not therefore be regarded as a means of providing impact insulation but of providing a small addition to some other insulation technique.

3.3 Floor Coverings

As has been mentioned, the obvious solution to the impact transmission problem lies in reducing the impact effect on the main structure. This can be achieved very successfully by covering the floor with a resilient layer such as carpeting or rubber tiling. The action of the resilient layer is to cushion the blow, thereby reducing the energy transmitted to the structure. Floor coverings are most effective in reducing the higher frequencies of the impact noise.

Unfortunately, the occupants of flats cannot be relied upon to provide adequate covering for the floors, and in some cases the stairs, and it is unrealistic for the designer to assume that they will.

In other cases, such as hospitals, schools and offices it is not, for a variety of reasons, the general practice to provide very resilient floor coverings and some form of built in resilience with a rigid covering capable of carrying the normal loads may be required. Such a construction is known as a floating floor.

3.4 Floating Floors

There are several types of floating floor based on both concrete and timber but the principle of operation is the same for them all. The floor which receives the impacts is supported above the main structure by either a continuous resilient layer or a series of separate rubber mounting blocks. The system is then analogous to that in Figure 3.1 and the insulation curves for different damping quantities are shown in Figure 3.2.

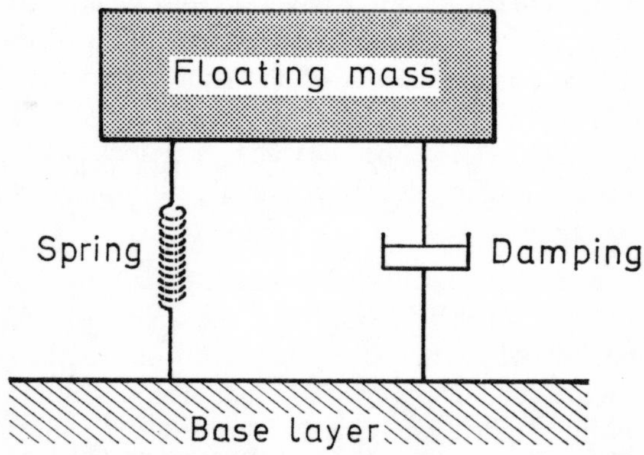

Figure 3.1. Principle of the floating floor.

It can be seen from Figure 3.2 that unless the damping is very high the insulation at the resonant frequency of the construction is lower than if the floating floor had been omitted. To avoid this problem the resonant frequency must be chosen to be very low, preferably lower than 20 Hz. This is achieved by having a high floating mass such as 50 mm (2 in) of concrete screed laid on a very resilient support such as 50 mm glass fibre quilting. Above the resonant frequency the insulation increases at a maximum rate of 12 dB/doubling of frequency when the damping is negligible and at a lower rate as the damping is increased.

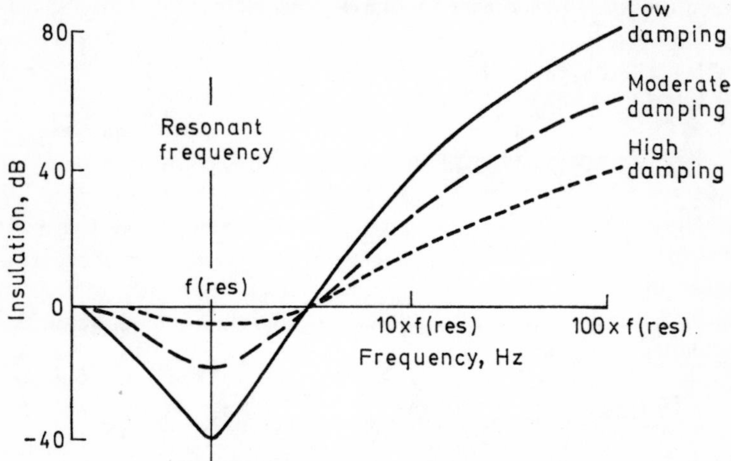

Figure 3.2. Insulation increase due to a floating floor.

Three types of floating floor construction are shown in Figure 3.3. Type A consists of an elastic insulating layer resting on the solid reinforced concrete structure. The resilient layer is covered with roofing felt and a lightly reinforced concrete floor is poured over the felt. Care must be taken to ensure that the floating concrete floor does not make contact with either the solid floor or the walls and skill is required to produce a floating floor which will not crack. A further covering of felt cardboard and linoleum is optional, giving slightly extra insulation especially in the middle frequency range.

Type B comprises a solid reinforced concrete base with distributed elastic supports. The area around the supports is filled with porous absorbing material either loose or in mat form. A plate of sufficient rigidity to carry the floating floor is placed over the mounts and the concrete poured on top. Care must be taken to ensure that there are no direct sound-bridges linking the floating floor to the walls or solid floor.

Type C is constructed of timber, the main joists carrying the ceiling of either plaster on lath or plaster finished plasterboard. A resilient quilt is laid over the joists and the floating floor of tongued and grooved boards on battens rests on the quilt. To provide adequate airborne sound insulation, sand or slag-wool lagging is laid between the joists to increase the mass.

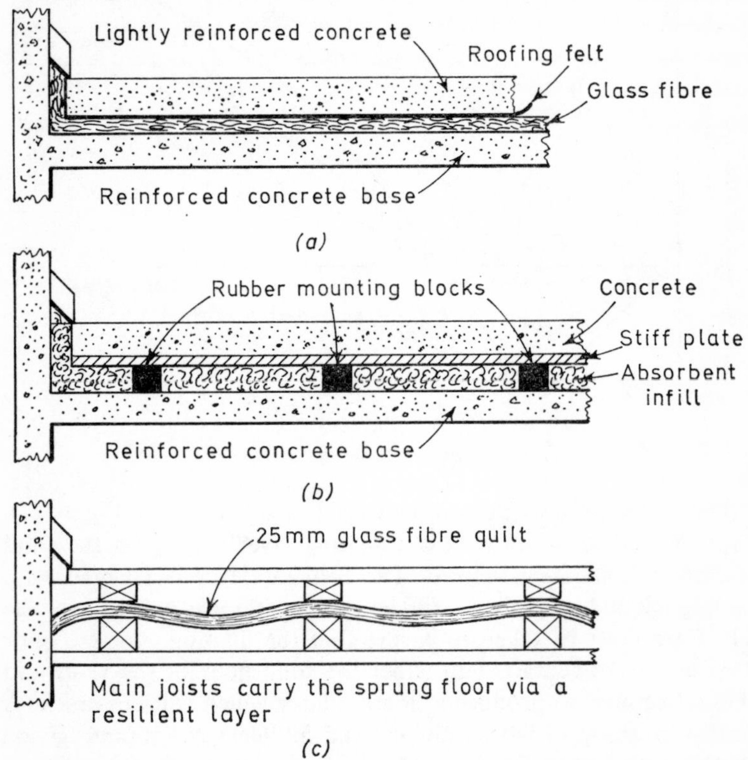

Figure 3.3 Three floating floor constructions.

All three constructions give top grade airborne and impact sound insulation.

3.5 Suspended Ceilings

These can improve the insulation against both airborne and impact noise by an amount depending on the weight of the suspended ceiling and the degree of rigidity with which the ceiling is attached to the structural floor. Such ceilings reduce the noise level only in the room where they are used, by decreasing the radiation from the ceiling treated; no reduction is obtained in other rooms in the building having an ordinary ceiling. Furthermore, they do not reduce the radiation of sound from the side walls which results from flanking transmission. To contribute substantially to the sound insulating quality of a floor the suspended ceiling must possess the following characteristics (Figure 3.4).

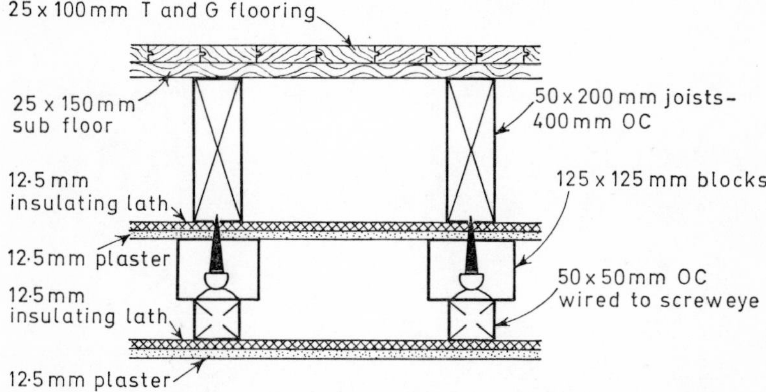

Figure 3.4. Suspended ceiling arrangement.

(*a*) The ceiling membrane must not weigh less than 25 kg m^{-2} (5 lb ft^{-2}). If an absorbent blanket (mineral or glass wool) is used in the air space above the ceiling, the weight of the ceiling membrane can be reduced.

(*b*) The ceiling membrane should not be too rigid.

(*c*) Direct paths of noise transmission through the ceiling must be avoided by the use of a solid and air-tight membrane, i.e. porous acoustic tiles are valueless.

(*d*) Gaps between ceiling and surrounding structure must be sealed to avoid noise penetration through direct air paths.

(*e*) The number of points of suspension from the structural floor above should be kept to a minimum; the use of resilient hangers is preferred to rigid ones.

(*f*) The air space between the ceiling membrane and the structural floor is increased to a reasonable maximum. As the gap increases the mass per unit area of the ceiling can be reduced.

It is important to realise that the suspended ceilings normally used in practice provide very little impact sound insulation and ceilings must be specially designed for this purpose. Consequently, suspended ceilings are not recommended as a means of obtaining high insulation in new buildings.

CHAPTER 4

Measurement of Insulation

4.1 Introduction

The sound transmission coefficient of a partition (wall, floor, roof, etc.) refers to the ratio of the transmitted sound energy to the incident sound energy. However, it is not possible to measure directly the energy in a sound wave. It is necessary to measure the sound pressure and relate the energy to this from a knowledge of the sound field characteristics. This latter requirement, in turn, demands the use of either anechoic or reverberant conditions so that simplifications make the mathematics more tenable. Experiments which involve a knowledge of sound energies, such as sound transmission problems, have therefore developed in a variety of ways using different conditions and measurement techniques. Not surprisingly the various methods have differed sufficiently to make direct comparisons of the results difficult and sometimes meaningless.

In 1955 proposals were made to the International Standards Organisation for standardised sound insulation techniques. By 1960 the draft recommendations had been accepted by the majority of the Organisation's member bodies and were published as the ISO Recommendation R140, Field and Laboratory Measurements of Airborne and Impact Sound Transmission.

The recommendation defines carefully the techniques and conditions required for the above measurements to be made to International Standards and can be obtained from the British Standards Institution. The following notes are an introduction to the recommended techniques and conditions.

4.2 Airborne Sound Transmission: Laboratory Measurements

The wall specimen under test is inserted in an opening between two reverberant rooms which should have minimum volumes of 50 m^3 (1750 ft^3) and preferably greater than 100 m^3 (3500 ft^3). The shape of the test rooms should be chosen to give an adequately diffuse sound field with the sound generation equipment used. Where the sound diffusion is inadequate some improvement can be made by the use of reflecting panels randomly positioned throughout the room. It is recommended that the area of the test wall should be approximately 10 m^2 (100 ft^2) with a minimum dimension of at least 2·5 m (8 ft) and the mounting conditions should be as near to practical conditions as possible. The flanking transmission should be negligible compared with the direct transmission through the specimen, thus for tests on walls of high insulation the two rooms should be vibration isolated from each other and of a substantial construction.

The sound used may be either warble tone or white noise generated by one or more loudspeakers and the measurements are made at $\frac{1}{3}$ octave intervals commencing at 100 Hz and ending at 3150 Hz. If warble tone is used the Recommendation specifies the frequency variation and if white noise is used the generator output in each band is filtered through specified filters as is the monitored noise. The sound pressure levels are measured at several positions in both rooms. This is necessary because the sound field is never completely diffuse, and the average sound pressure level throughout the room must be obtained. The average pressure level should, strictly, be obtained from the arithmetic mean of the energy at each position at each frequency. However, it is usually sufficient to take the arithmetic mean of the decibel values.

The laboratory measured sound insulation of a partition is usually quoted in terms of the sound reduction index (R). It is equal to the number of decibels by which sound energy incident on the partition is reduced in transmission through it. The numerical value of R depends on the construction and mounting of the partition only; it is independent of the acoustical properties of the two spaces separated by the partition. However, the level in the receiving room will depend not only on the insulation provided by the partition but also on the area of the partition and on the amount of absorption in the receiving room.

If the average sound pressure level in the sending room is L_1 and in the receiving room is L_2 then the *sound reduction* is given by

$$\text{Sound reduction} = L_1 - L_2$$

This must be corrected to give the *sound reduction index* R, as L_2 is dependent on the conditions in the receiving room and on the area of the partition.

The S.R.I. is given by $R = L_1 - L_2 + 10 \log S - 10 \log A$

Where S is the area in square metres (or square feet) of the partition being measured and A is the absorption in square metres (or square feet) in the receiving room.

R is calculated for every frequency and the results are presented in graphical form.

Note: To calculate R, A must be measured at each frequency, usually by measuring the reverberation time.

4.3 Airborne Sound Transmission: Field Measurement

There are two minor differences between laboratory and field measurements.

(*a*) The size of the partition. If the size of the partition being measured in the field is very different from the laboratory specimen then its sound insulation behaviour will be different.

(*b*) The edge-fixing conditions between field and laboratory will probably be different, and this may have some effect, depending on the construction.

However, the major difference between laboratory and field measurements is that in the laboratory indirect transmission is always negligible but in the field it may predominate. Hence, it is not usually possible to measure the sound reduction index of a partition in the field. Instead the normalised level difference (D_n) is measured. The test technique is basically the same as for the laboratory measurements except that usually more loudspeakers are required to produce a reasonably diffuse sound field in both rooms.

The average sound pressure levels are measured in both the sending and receiving rooms and the difference is then corrected to allow for a change in the conditions of the receiving room—two methods of correction may be used.

The first method of correction is to use a reference reverberation time for the receiving room. The corrected value of insulation in decibels is then given by

$$D_n = L_1 - L_2 + 10 \log T - 10 \log T_{ref}$$

where L_1 and L_2 are the average sound pressure levels in the sending and receiving rooms, T is the measured reverberation time of the receiving room and T_{ref} is the reference reverberation time.

The second method is to use a reference absorption of 10 m^2 (100 ft^2) in the receiving room. The corrected value of insulation is then given by:

$$D_n = L_1 - L_2 + 10 \log A_{ref} - 10 \log A$$

Where A is the measured absorption in the receiving room.

It should be noted that L_1 and L_2 are again measured in $\frac{1}{3}$ octave intervals from 100 Hz to 3150 Hz. Also that T and A are determined for every measured frequency.

In dwellings it has been found that the reverberation time of furnished living rooms is usually about 0·5 sec, independent of their volume. This means that, if the measured values are corrected to a standard reverberation time of 0·5 sec, the corrected measured values will be close to the values existing in practice when the rooms are furnished and occupied.

The normalised level difference in decibels is then given by:

$$D_n = L_1 - L_2 + 10 \log T - 10 \log 0{\cdot}5.$$

Again it is preferable to present the results in graphical form.

Example

At 250 Hz the sound pressure levels at six positions in the same room were 99, 100, 95, 95, 97 and 98 dB.

$$\text{Hence} \quad L_1 = \frac{99 + 100 + 95 + 95 + 97 + 98}{6} = 97$$

In the receiving rooms the levels were 60, 55, 59, 59, 63 and 59 dB.

$$\text{Hence} \quad L_2 = \frac{60 + 55 + 59 + 59 + 63 + 59}{6} = 59$$

The measured reverberation time in the receiving room is 1·53 sec.

$\therefore$ $D_n = 97 - 59 + 10 \log 1{\cdot}53 - 10 \log 0{\cdot}5 = 38 + 5 = 43$ dB.

At 250 Hz the normalised level difference between the sending and receiving room is 43 dB.

4.4 Impact Sound Transmission: Laboratory Measurements

Impact sound insulation is measured using a standard "impact" machine. This produces impacts of standard energy on the surface of the floor being measured, and the resulting sound pressures in the room below are measured (Figures 4.1 and 4.2). It should be noted

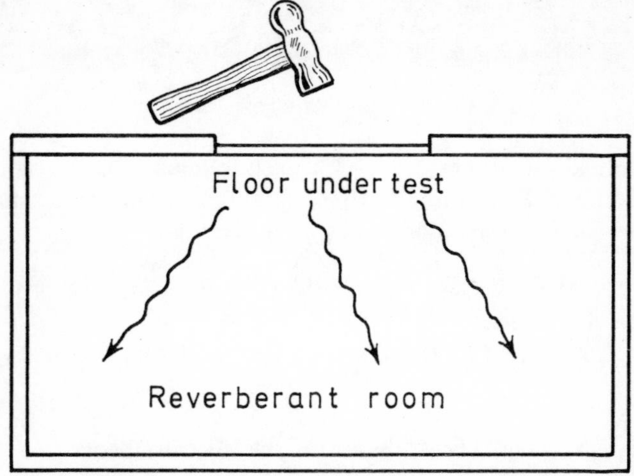

Figure 4.1. Symbolic diagram of floor impact test.

that unlike airborne sound insulation measurements which are only relative, an absolute measure of sound pressure is now required, needing a calibrated microphone.

The pressure-level measurements in the receiving room are made at several positions and the results averaged. Measurements are made at octave intervals beginning at 125 Hz and ending at 3150 Hz.

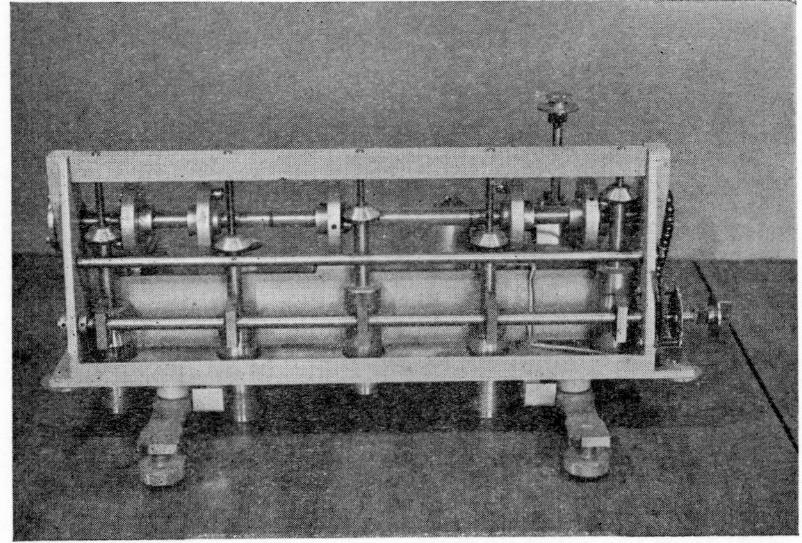

Figure 4.2. Standard tapping machine.

The average sound pressure level in each frequency band is adjusted to a reference absorption of 10 m² (100 ft²). Thus the normalised impact sound transmission level (L_n) is given by:

$$L_n = L + 10 \log A - 10 \log A_{ref}$$

where L is the measured level and A is the measured absorption in the receiving room.

A high measured L_n thus means a low insulation and vice versa.

4.5 Impact Sound Transmission: Field Measurements

The differences between laboratory and field tests tend to be greater for impact insulation than for airborne insulation. This is because not only is the indirect transmission different but the size of the floor and the edge conditions have a greater effect. Further, the laboratory method of correction makes no allowance for size. Laboratory measurement can solve many problems of importance, but the results are frequently of less interest in building practice.

For field measurement the impact sound is generated using the standard impact generator. The average sound pressure levels are again measured and corrected to allow for the condition of the receiving room. The levels can either be corrected to a standard absorption of 10 m^2 or to a standard reverberation time.

The normalised impact sound transmission level is thus given by:

$$L_n = L + 10 \log A - 10 \log A_{ref}$$

or

$$L_n = L + 10 \log T_{ref} - 10 \log T$$

where L is the measured level, A is the measured absorption of the receiving room and T the measured reverberation time.

The grading system for impact insulation between dwellings applies to measurements corrected to a reverberation time of 0·5 sec.

Measurements are made in the octave bands 125 Hz to 4000 Hz and the results are presented graphically.

The impact insulation between a particular floor and any room in the building can be measured in the same way; the test is not restricted solely to measurements from a floor to the room immediately below.

CHAPTER 5

Ventilation Systems

5.1 Introduction

The installation of ventilation plant introduces two main problems.

(1) The fan and motor produce noise and vibration which can be disturbing even in rooms not served by the system.

(2) The ventilation ducts can provide short circuits to the insulation between rooms and to insulation from external noise.

The effects of vibration can be reduced by siting the plant in an area where the disturbance is not very critical. In addition the fan and motor can be mounted on resilient anti-vibration pads which, in effect, stop energy from being transmitted into the building structure. Similarly, canvas connectors should be used to isolate the fan housing from the ductwork. Noise radiated directly from the fan and motor can be contained within the plant room by normal airborne sound insulation principles.

This leaves the airborne fan noise which, unless means are taken to prevent it, will be channelled along the ducts and into the rooms. If the ducts serve two or more rooms it is also possible for noise from one room to travel to others by that route. Figure 5.1 illustrates a typical ventilating system in which steps have been taken to avoid some of these difficulties.

Other problems which may arise are:

(1) If the ducts have very thin walls, noise may be transmitted through them.

(2) High speed flows 15 m sec^{-1} (3000 ft min^{-1} or greater) give rise to a significant amount of self-noise. This is related to turbulence and is aggravated by bends, obstructions, guide vanes, etc.

(3) Outlet grille noise which is worse with guide vanes than with open mesh and increases with flow velocity.

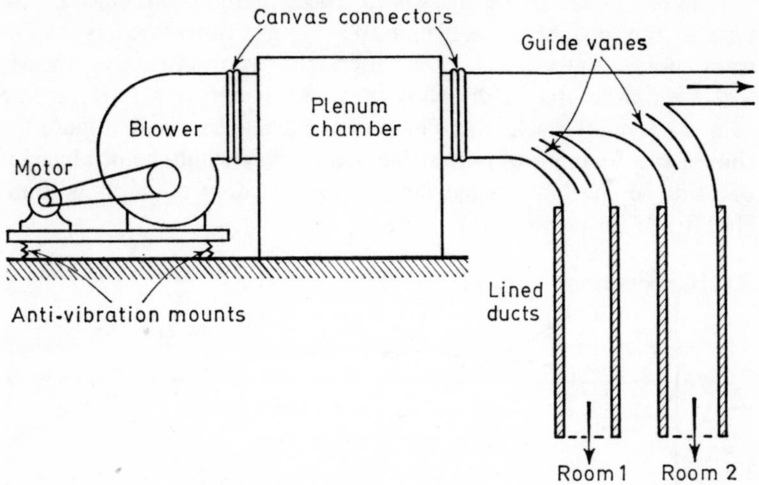

Figure 5.1. Typical ventilation system

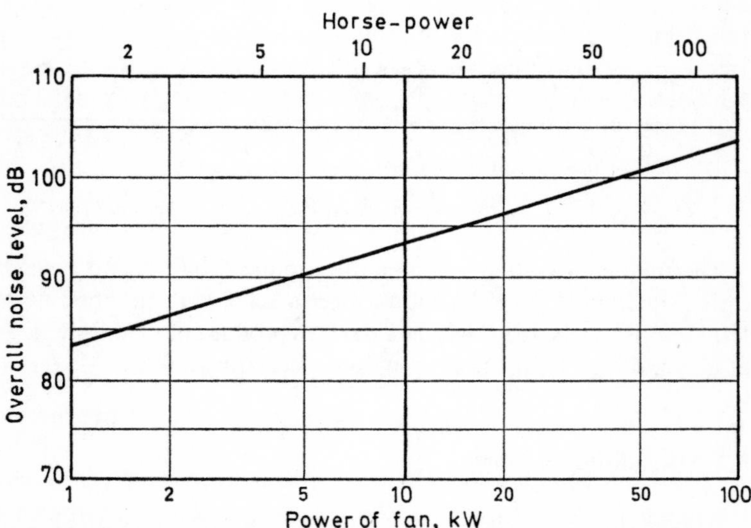

Figure 5.2. Noise level in standard room as a function of fan power.

5.2 Fan Noise

Fans can be either the high-speed (1000 r.p.m. upwards) axial flow type or the low-speed (less than 1000 r.p.m.) centrifugal type. The total noise output of both increases approximately linearly with the power; that is, doubling the power results in a 3 dB increase in noise level. To give some idea of the levels involved, Figure 5.2 shows, as a function of power, the noise which would be produced if one side of the fan opened directly into a room containing 10 m² (100 ft²) of absorption.

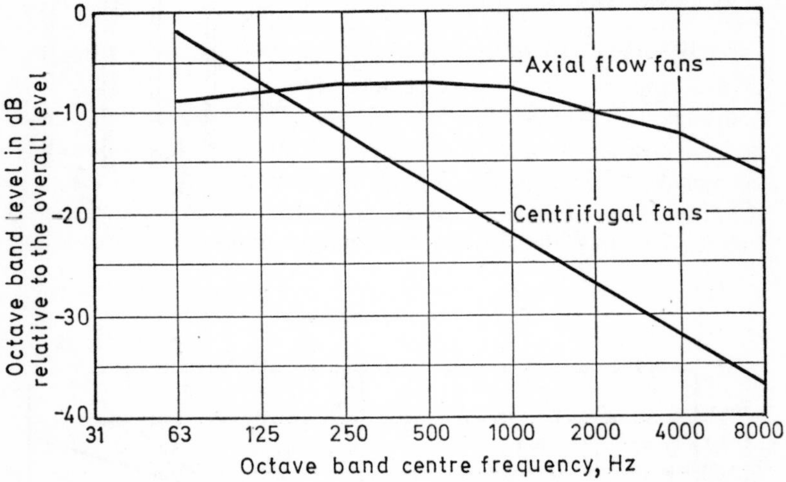

Figure 5.3. Noise spectra of axial flow and centrifugal fans.

The frequency content of the noise produced by the two types is rather different. Figure 5.3 shows spectra relative to the total noise level and it can be seen that the axial fans produce more noise energy at mid and high frequencies than do the centrifugal fans.

5.3 Attenuation in Ducts

Attenuation in straight unlined ducts is usually very small and except in very long ducts is negligible. Some attenuation is given by bends and changes in cross-sectional area but again this is often small unless the bends are numerous and sharp and the changes in cross-sectional area are large. Even when a duct enters a room, for example, the attenuation is unlikely to be more than 2 or 3 dB. This type of attenuation is due to reflection of sound energy back towards the source.

To obtain any appreciable reduction some sort of sound absorption must be employed. One of the simplest ways is to line the duct walls with a sound absorbent material. These linings should possess the following properties.

(*a*) High absorption coefficient.

(*b*) Smooth surface for low air friction.

(*c*) Adequate strength to resist disintegration due to air stream.

(*d*) Odourless, fire, rot and vermin proof.

Provided that the duct is not greater than about 0·3 m² in area and the ratio of the sides does not exceed about 2:1, the attenuation per unit length is related by a fairly simple empirical relationship to the absorption coefficient, α, of the lining material. This is shown in Figure 5.4 and it is quite feasible to obtain an attenuation

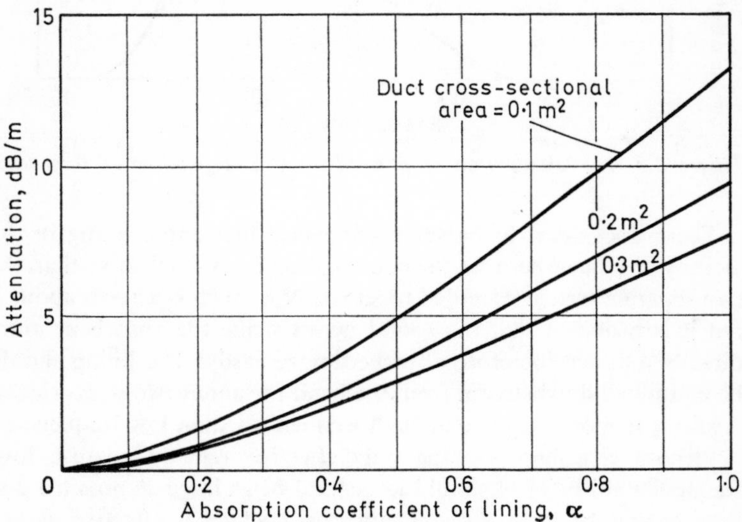

Figure 5.4. The attenuation of a lined duct as a function of the lining absorption coefficients.

of 6 to 10 dB per metre (2 to 3 dB per ft). Evidently the attenuation is going to be a function of frequency because the absorption coefficient is a function of frequency and in general it is easier to attenuate high frequency than low frequency noise. Figure 5.5 shows a typical spectrum of absorption against frequency for a lined duct.

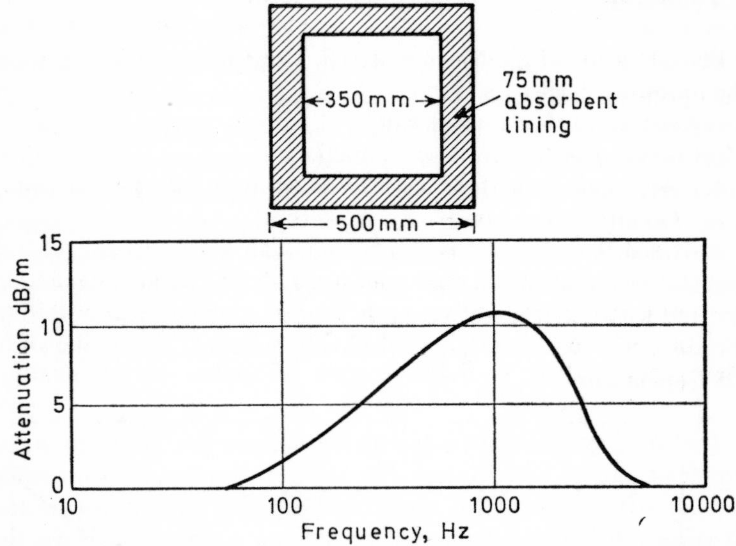

Figure 5.5. The attenuation of a lined duct as a function of frequency.

There does seem to be some advantage in using a lining in the vicinity of a bend because the combination is more effective than the two separate measures added together. Not much is known about it but it is probable that the sound waves strike the absorbent more directly and are therefore absorbed more easily. The lining should be continued downstream from the bend for about two duct-widths. Again it is more effective at high frequencies than low frequencies.

Plenum chambers are the only effective remedy against low-frequency noise. They should in general be as large as possible and they should be lined with an absorbent which is effective at low frequencies, e.g. a thick pad of glass fibre or absorbent plastic foam. Additional absorption can be obtained by including baffles in the chamber. The cavities which occur naturally in modern buildings can often be used as plenum chambers, although the sides must obviously provide sufficient insulation for them not to be bothersome to adjacent rooms.

5.4 Silencers

Commercially available silencers are normally based on the lined duct principle, although of course a considerable amount of research and development has gone into them to produce maximum efficiency. Usually the cross-section is sub-divided by absorbent splitters which effectively increase the duct perimeter and, therefore, the absorbent area. The attenuation is still a function of frequency usually reaching a maximum at mid-frequencies. If, however, a high attenuation is required in a short distance a "package" unit is probably the answer.

5.5 Conclusion

The firms specialising in ventilation systems are aware of noise problems and of how to deal with them. In general, it is probably most satisfactory to specify the acceptable noise levels in rooms and to specify the insulation required between rooms, and leave the complete design to a commercial organisation capable of designing and installing the system.

CHAPTER 6

Traffic Noise

6.1 Introduction

Traffic is by far the greatest contributor to town noise. One can obviously think of aeroplanes and trains or demolitions and road-work but, though disturbing, they are local and usually short-term. They must still be taken into account if buildings are to be sited near an airport or a railway, but traffic noise is probably of more general interest. Indeed, it is difficult to site a building anywhere but in the vicinity of a road.

6.2 Noise in Built-up Areas

An extensive survey in the London area gives some idea of the noise climate which already exists in the vicinity of roads other than motorways. Table 6.1 is reproduced from the Wilson Committee Report on Noise. It should be noted that the higher figures are exceeded for 10% of the time and in fact peak levels can be as much as 10 dB higher than the 10% levels.

The levels shown in Table 6.1 are likely to remain fairly static. It is possible that they will creep upwards by a few dB as the volume of traffic increases but legislation will limit the production of noise by vehicles to that which they have been producing in recent years.

6.3 Noise from Motorways

A newer, and perhaps more serious, threat is from urban motor-

TABLE 6.1

RANGE OF NOISE LEVELS AT LOCATIONS IN WHICH TRAFFIC NOISE PREDOMINATES

Group	*Location*	*Noise climate in dB (A)**	
		Day 8 *am–6 pm*	*Night* 1 *am–6 am*
A	Arterial roads with many heavy vehicles and buses (kerbside)	80–68	70–50
B	(1) Major roads with heavy traffic and buses (2) Side roads within 15–20 m of roads in groups A or B (1) above	75–63	61–49
C	(1) Main residential roads (2) Side roads within 15–20 m of heavy traffic routes (3) Courtyards of blocks of flats, screened from direct view of heavy traffic	70–60	55–44
D	Residential roads with local traffic only	65–56	53–45
E	(1) Minor roads (2) Gardens of houses with traffic routes more than 100 m distant	60–51	49–43
F	Parks, courtyards, gardens in residential areas well away from traffic routes	55–50	46–41
G	Places of few local noises and only very distant traffic noise	50–47	43–40

* By noise climate is meant the range of noise level recorded for 80% of the time. For 10% of the time the noise was louder than the upper figure and for 10% of the time it was less than the lower figure.

ways which are being planned and built in many of our towns. The noise aspect is usually far down the list of priorities and it is often not until the road is open to traffic that people discover just how

TABLE 6.2

MAXIMUM PERMISSIBLE NOISE LEVELS FROM ROAD VEHICLES—JULY, 1968

Type of vehicle	*Noise level in dB (A)*
M/C under 50 c.c.	80
M/C over 50 c.c.	90
Heavy vehicles	92
Buses	92
Mini-buses	87
Cars	88

disturbing it is. Quite a lot is already known about the noise produced by moving traffic on both restricted and unrestricted roads and in each case the roadside level can be directly related to vehicle density. Levels are shown in Figure 6.1. Neglecting, for the moment, the shielding effect of barriers and cutting walls the noise diminishes as a function of distance and the type of ground over which it travels. This is illustrated in Figure 6.2. Close to the motorway the noise level diminishes fairly rapidly with distance but as the distance becomes greater there is apparently a law of diminishing returns. It is clear then that if the noise level at a distance of, say, 100 m from the road is unsatisfactorily high the solution is not to go farther from the road but to provide some extra protection.

6.4 Shielding

The answer is to improve the insulation of the facade, perhaps by double glazing, or even to avoid having windows overlooking the road. If the problem is more communal in nature, i.e. if there are many buildings involved, the solution is more likely to be found in providing some sort of barrier alongside the road or even by sinking

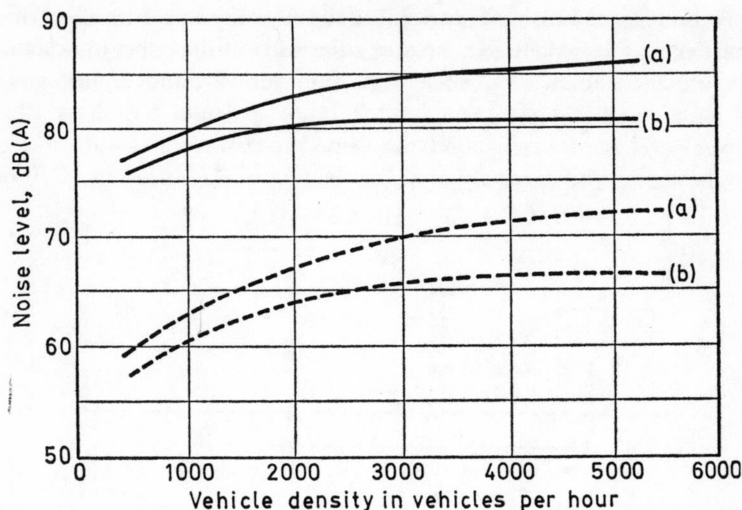

Figure 6.1. Roadside noise levels: (a) for unrestricted roads, (b) for roads subject to a 50 k.p.h. speed limit. ——— *10% noise levels,* ...... *90% noise levels.*

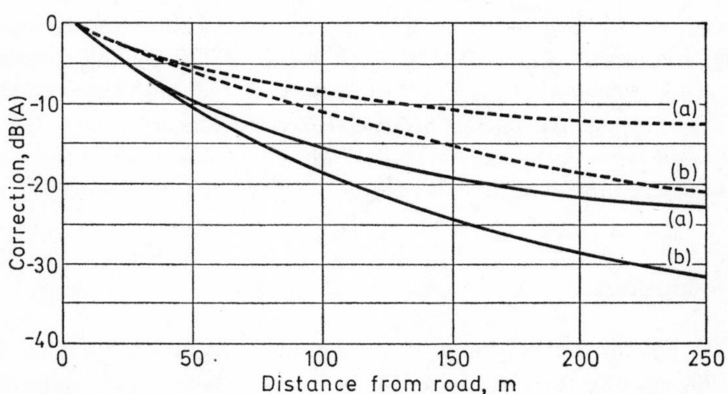

Figure 6.2. Effect of distance from road on noise level: (a) over a hard surface or in free space, (b) at ground-floor level over uncut grass. ——— *10% noise levels,* ...... *90% noise levels.*

the road into a cutting. Figure 6.3 shows how the shielding effect of a barrier may be calculated. To give some idea of the effect in practice Figure 6.4 shows 10% noise level contours around a motorway situated on flat open land, while Figures 6.5 and 6.6 show 10% noise level contours around the same motorway, first with a 3 m high barrier and secondly in a 3 m deep cutting. Shielding of 10 or

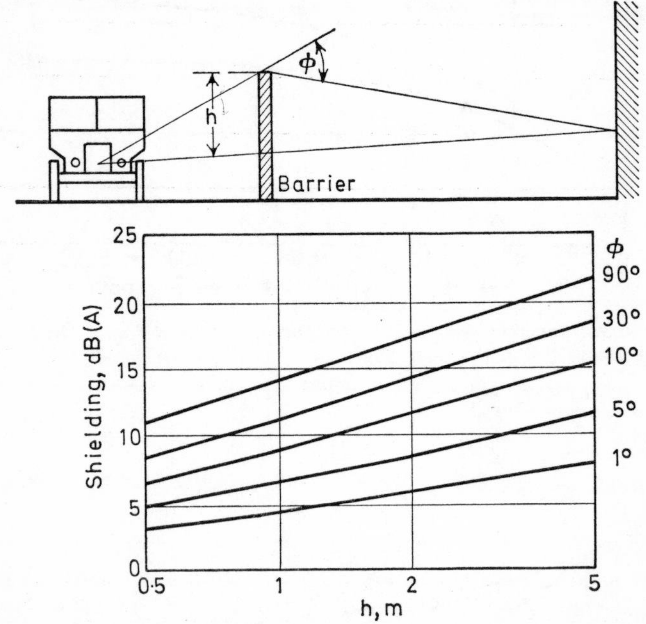

Figure 6.3. Shielding effect of a barrier.

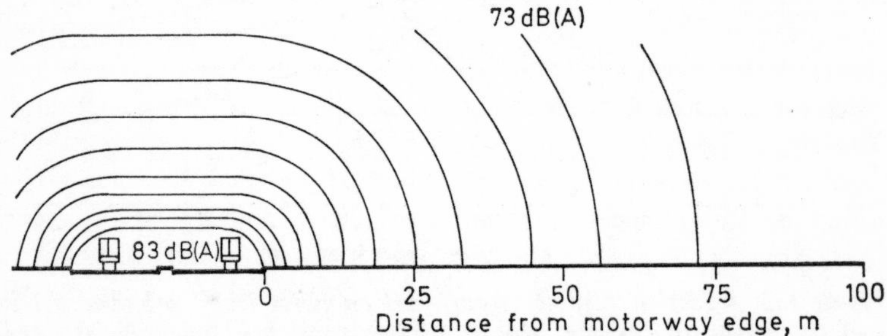

Figure 6.4. Noise contours around a surface motorway.

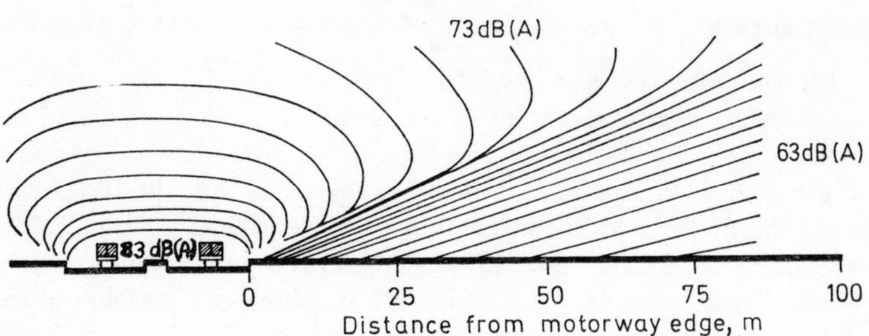

Figure 6.5. Noise contours around a surface motorway with 3m high barriers.

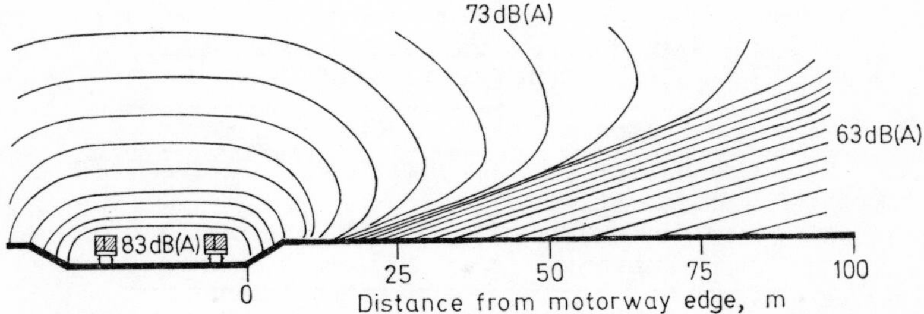

Figure 6.6. Noise contours around a sub-surface motorway with 3m high banking.

15 dB can be obtained for the lower storeys but if the adjacent buildings are very tall then it is necessary to build higher barriers or even partially enclose the roadway by horizontal extensions. If it is essential to build an elevated roadway, as sometimes happens, no ground absorption can occur and the noise contours, if there are no barriers, will resemble those in Figure 6.4, the only difference being that the road surface itself can shield the lower storeys of nearby buildings. To reduce noise levels in the vicinity of elevated roads again requires barriers, unsightly though they may be. The barriers must be fairly substantial and should, of course, be completely unperforated. There may be occasions when the inside of the barrier should be covered with some absorbent material to avoid reflection on to buildings on the other side of the road, but this of course depends on the location of the particular road.

6.5 Criteria

Information has been given in terms of the 10% and the 90% noise levels. The former may be thought of as corresponding to the louder, more annoying contribution, while the latter corresponds to the general background noise. It is suggested that the 10% noise levels should be used in conjunction with the criteria given in Tables 6.3 and 6.4. Current research work has produced a new unit, the Traffic Noise Index (T.N.I.), which is a weighted combination of the 10% and the 90% noise levels but as yet this has only been shown to be satisfactory under limited conditions, and no general criteria in terms of the T.N.I. are available.

TABLE 6.3
RECOMMENDED INTERNAL 10% NOISE LEVELS FOR DWELLINGS

Situation	*Day*	*Night*
Country areas	40 dB A	30 dB A
Suburban areas (away from main traffic)	45 dB A	35 dB A
Busy urban areas	50 dB A	35 dB A

TABLE 6.4
RECOMMENDED INTERNAL 10% NOISE LEVELS FOR BUILDINGS OTHER THAN DWELLINGS

Type of room	*American criteria in dB (A)*	*European criteria in dB (A)*
Broadcasting studio	25–30	20
Concert hall	25–30	25
Legitimate theatre (500 seats)	30–35	25
Classroom	35	30
Music room	35	30
TV studio	35	30
Courtroom	30–35	35
Library	40–45	35
Cinema	40	35
Hospital	40	35
Church	35	35
Restaurant	55	50
Private office	40–45	40
General office	60–65	55

CHAPTER 7

Absorption of Sound

7.1 Introduction

If nothing can be done to reduce the sound power output of a source, the sound energy it emits must be absorbed (that is, converted into heat energy) sooner or later. The interposition of barriers merely postpones or relocates the problem. In addition, the "acoustic ambience" of rooms is dependent on the absorptive properties of the walls. An understanding of absorption mechanisms is therefore important both from the point of view of noise control and for an understanding of how room acoustics may be controlled. This chapter is therefore concerned with a discussion of absorption mechanisms and the properties of absorbing surfaces.

7.2 The Inverse Square Law

The movement of air and the change in its pressure produced when a sound wave passes are forms of energy: air through which sound passes is energised. So long as a sound source persists, it is sending out energy into the air around it. In the open air the sound spreads out as it travels away from the source so that the intensity (energy crossing unit area per second) automatically decreases with distance, because the available energy is being spread over greater areas. For this reason a point sound source of power W watts which radiates spherically should produce an intensity of $W/4\pi r^2$ watts m^{-2} at a distance of r metres, or a line source of strength W watts m^{-1} should produce an intensity of $W/2\pi r$ watts m^{-2} at r metres, provided r is smaller than the length of the source. Written in terms of intensity level, these become

I.L. = $(10 \log_{10} W - 20 \log_{10} r + 120 - 10 \log_{10} 4\pi)$ dB, point source and I.L. = $(10 \log_{10} W - 10 \log_{10} r + 120 - 10 \log_{10} 2\pi)$ dB, line source

The first equation shows that there is a 6 dB reduction in the intensity level due to a point source whenever the distance from it is doubled. The corresponding fall for distance doubling in the case of the line source is shown by the second equation to be only 3 dB.

The interposition of barriers will reduce the amount of sound energy reaching a given point, by reflecting some of the sound back towards the source and beyond provided the dimensions of the barriers are large compared to the wavelength of the sound. However, neither removal of the source to a greater distance, nor the interposition of barriers reduces the amount of energy in the air; there is merely a spreading or redirection of the sound. True absorption only takes place when the energy of the air imparted by the sound wave is converted irreversibly into heat energy. This occurs to some extent while the sound passes through the air (air absorption) and, more important, whenever it is reflected from a surface (surface absorption).

7.3 Air Absorption

Air is not perfectly elastic, so that some of the sound energy is converted into heat energy in the course of alternately squeezing and expanding it. The effect is a composite of a number of processes which are all strongly dependent on the frequency of the wave, so that air absorption is negligible at low frequencies. At 2000 Hz there is a reduction in intensity level of 0·008 dB for every metre of travel. At 4000 Hz this becomes 0·03 dB per m and at 8000 Hz, 0·08 dB per m. (These figures are for air with a relative humidity of 60%. The absorption at first increases as the relative humidity decreases, but then passes through a maximum, before falling to much lower values in dry air.)

Care must be taken to distinguish between this "internal"

absorption and the scattering produced by wind and temperature gradients. This type of attenuation predominates out of doors, but often varies with the weather. It again involves redirection of the energy (as with barriers) rather than true absorption.

7.4 Boundary Absorption

When the pressure changes and movement of air induced by a sound wave take place close to a surface, the surface may move slightly under the influence of the pressure changes, or, if it is porous, air may flow in and out of it. The movement of the surface, or of the air in its pores, will involve the expenditure of energy against the viscosity or friction in the surface. In this way energy is extracted from the sound wave, so that the reflected wave is weaker than the incident wave. Strong reflection will only take place if the surface is impervious and rigid. At very high frequencies even an impervious rigid surface absorbs sound to some extent due to heat exchange with the compressed (heated) and expanded (cooled) regions of gas.

In this way different surfaces absorb sound that falls on them to different degrees. The measure of the degree to which they do so is the *absorption coefficient* (α), which is simply the ratio of the sound energy which is *not* reflected from the surface to the sound energy incident on it. A perfect absorber (e.g. an open window) has an absorption coefficient of 1, while a perfect reflector has an absorption coefficient of 0. The absorption coefficient varies with angle of incidence, but a "statistical" average value can be quoted, representing conditions in a room. Alternatively the normal incidence value may be given.

Now, consider a room with perfectly reflecting walls and with negligible air absorption. A source sounding continuously in such a room would fill the room with sound energy to a steadily increasing density, that is, it would get steadily louder in the room, since there is nowhere else for the sound energy to go. This is prevented from occurring in real rooms by the absorption of sound at the walls and in the air which eventually brings about an equilibrium between the rate of radiation of sound and the rate at which it is absorbed. The absorbing surfaces take up a fixed fraction of the energy falling on them so the energy flowing round the room must rise to a sufficiently large value for the rate of extraction to equal the input rate. For this reason the total sound *absorbing power* in a room is very important, since it determines the equilibrium energy density in a room in which a noise source operates. The equilibrium reverberant intensity in a room can be calculated (in watts m^{-2}) from $I = 4W/A$, where W is again the power of the source in watts, and A is the total absorbing power in the room (in m^2), calculated most simply by adding together the product of absorption coefficient and area for all the surfaces in the room, including people and furniture.

In addition to the variation with angle already mentioned there is a very marked variation of absorption coefficient with frequency. In general, surfaces which absorb strongly in one frequency range will be less good in other frequency ranges. The variation with frequency as well as the actual absorption coefficients are both different for different types of surface, and this must be borne in mind when discussing the properties of surfaces.

7.5 Classification of Surfaces

Boundaries may be classified as follows:
(*a*) Impervious and rigid
(*b*) Porous and rigid
(*c*) Porous and flexible
(*d*) Impervious and flexible
In addition to these there are two types of "artificial" surface, designed for high absorption coefficient:
(*e*) Acoustic tiles
(*f*) Perforated plate multi-resonator
Single object absorbers must also be considered.
(*g*) Space absorbers
(*h*) Single resonators

7.6 Rigid Impervious Surfaces

As already mentioned, even when a surface is totally sealed some sound absorption takes place by heat exchange between the surface and the compressed and rarefied regions of gas close to it. In addition to this there will generally be some residual lack of rigidity which will allow the surface to yield very slightly under the impact of the sound wave and thus absorb some of its energy. Furthermore, the area of contact between sound wave and surface may be increased by roughness. (Care should be taken, however, to avoid confusing roughness with porosity.) Probably the maximum values of absorption coefficients from this source likely to be encountered are in stone churches, where they are generally low anyway:

125 Hz	250 Hz	500 Hz	1 kHz	2 kHz	4 kHz
0·01	0·01	0·02	0·02	0·03	0·04

In other buildings where there are large areas of more absorptive surfaces the "unavoidable" absorption of rigid impervious surfaces can often be neglected.

7.7 Rigid Porous Surfaces

A rigid porous material (such as a "Tyrolean" finish or sprayed asbestos fibre layer) is characterised by a number of variables.

(1) The *porosity*, which is the fraction of empty space within the material to its total volume.

(2) The *flow resistance*, which is a measure of the difficulty with which air can be blown through unit thickness of the sample (*see* Figure 7.1).

(3) The *structure factor*, which is a measure of the amount of dead space (such as cul-de-sac and pores running parallel to the surface) compared with pores which convey air through the material. A material with structure factor 1 has all its pores running straight from front to back (i.e. parallel with the direction of incident sound). Materials with more tortuous pores have structure factors greater than 1.

These factors are not independent of each other, but the development of the theory of porous absorbers is facilitated by their use. There are fairly straightforward experimental procedures for the measurement of porosity and flow resistance, but structure factor is best measured by applying the theory backwards, from measured values of absorption coefficient. An attempt to illustrate the differences in materials with porosity, flow resistance and structure factor varied separately is made in Figures 7.2, 7.3 and 7.4 respectively.

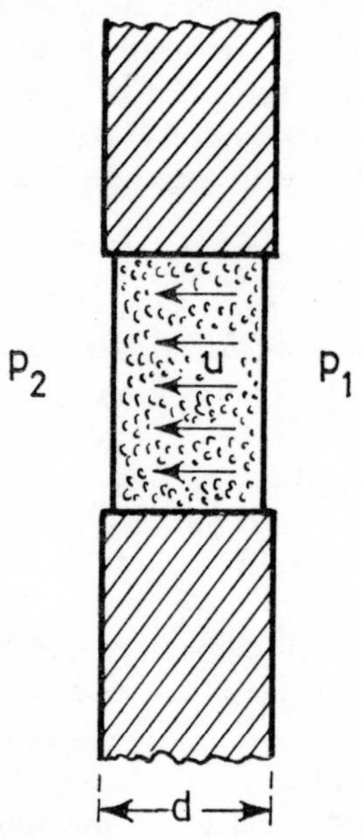

Figure 7.1. Definition of flow resistance, R. p = pressure in Nm^{-2}; d = sample thickness in m; u = flow velocity in m sec^{-1}.

$$R = \frac{\dfrac{p_1 - p_2}{d}}{u}$$

The way in which these factors, and others, affect the absorptive properties of porous materials can now be considered. Porosity is relatively unimportant, provided it is greater than about 70%. Larger values than this do not provide much improvement in absorptive properties, except in so far as the flow resistance is decreased. Flow resistance is of much greater significance, particularly in conjunction with the thickness of the layer of absorbent in front of a rigid wall.

Sound travelling through the air is said to experience a "characteristic impedance". This does not refer to the air absorption discussed above which is quite separate and much smaller, but

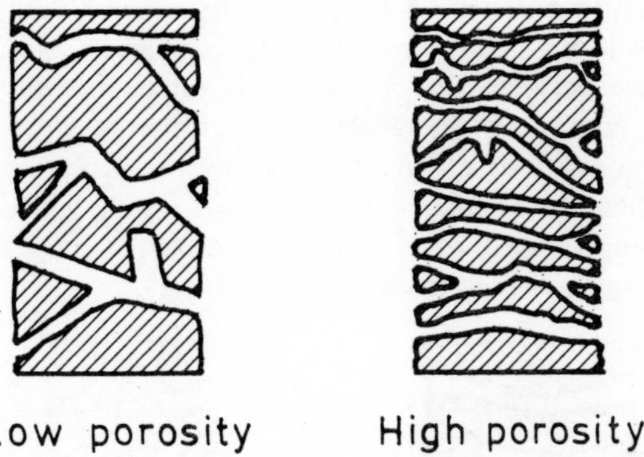

Figure 7.2. Variation of porosity. Flow resistance and structure factor held constant.

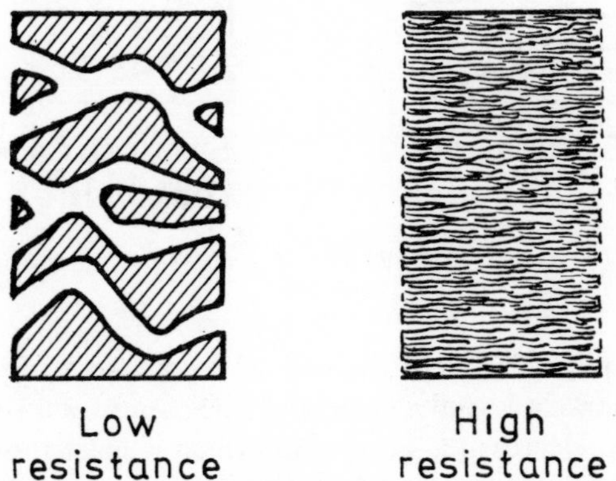

Figure 7.3. Variation of flow resistance. Porosity and structure factor held constant.

rather arises from the forces required to alternately squeeze and accelerate the air as the sound wave travels along. Energy stored in these processes is regained in the next half cycle. Now, whenever a sound wave encounters a boundary between two media (say between air and a porous absorber) some of the sound energy crosses into the second medium while the rest is reflected. The fraction that crosses will depend on how different are the respective values of characteristic impedance; if they are very similar then almost all the energy will cross over and little will be reflected, but if there is a big difference between the characteristic impedance of the media, most of the energy will be reflected. It is fairly easy to see that the higher the porosity and the lower the flow resistance of a porous absorber, then the smaller will be the mismatch between air and absorbent impedances, and so the higher will be the fraction of incident energy passing into the porous layer.

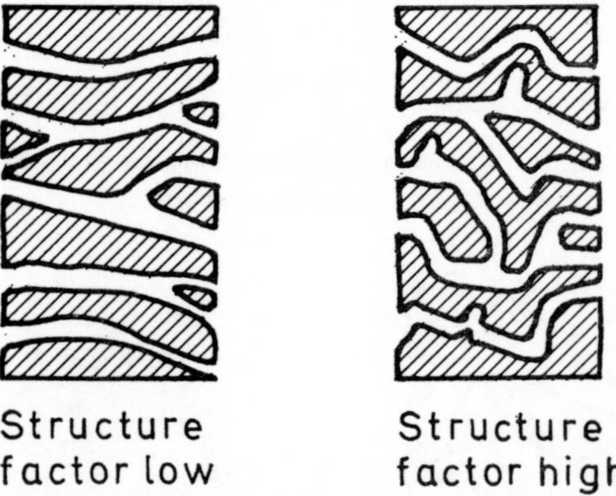

Figure 7.4. Variation of structure factor. Porosity and flow resistance held constant.

Unfortunately this is not the end of the sound wave. The wave that has penetrated into the layer continues to travel through it with its amplitude decreasing at a rate which depends on the flow resistance of the material, but now rapid energy absorption requires a *high* flow resistance. Sooner or later the wave will encounter a backing, rigid wall and most of the remaining energy will be reflected back through the layer. After further absorption some sound will finally re-emerge at the front of the layer and add to the energy sent back after the first encounter, by the impedance mismatch. This process is illustrated in Figures 7.5, 7.6, 7.7 and 7.8.

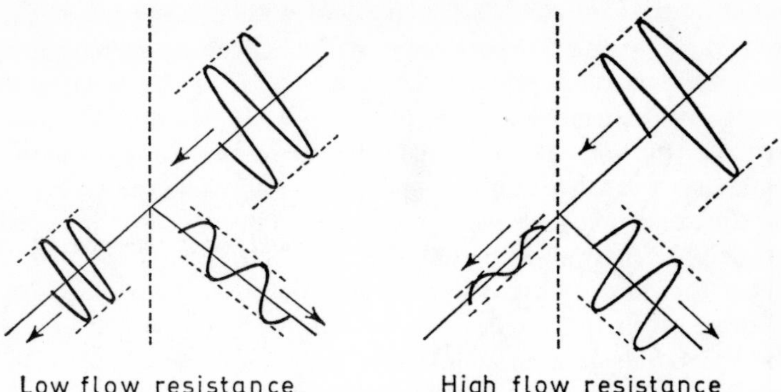

Figure 7.5. Sound wave crossing into porous absorber.

Because of this possibility it becomes necessary to compromise on the value of flow resistance for a *finite* layer. For each thickness there will be an optimum value of flow resistance which balances the requirements of easy penetration against satisfactory absorption in

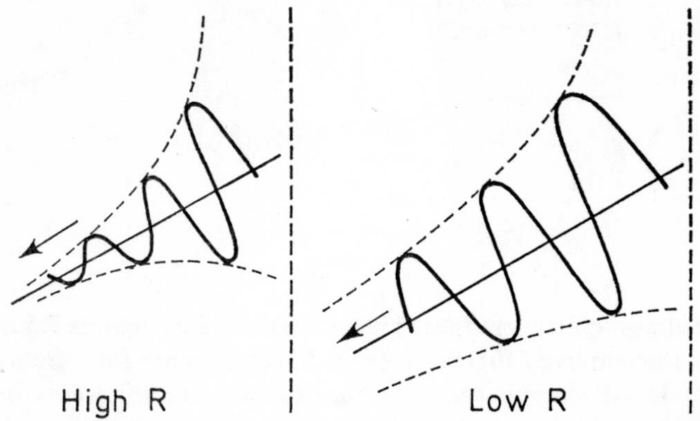

Figure 7.6. Sound wave penetrating absorbing layer.

the total path available (twice the thickness of the layer). At the same time, for material of a given flow resistance there is little point in increasing the thickness of the layer beyond the stage at which the re-emergent wave becomes negligible in comparison with the wave

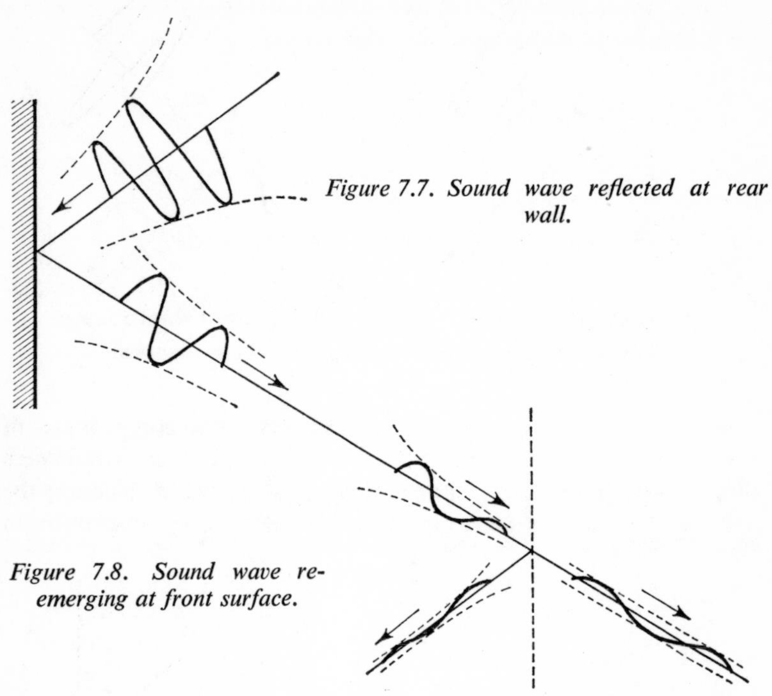

Figure 7.7. Sound wave reflected at rear wall.

Figure 7.8. Sound wave re-emerging at front surface.

reflected at the first interface. This is illustrated in Figures 7.5 to 7.8 and in the graphs of Figures 7.9 and 7.10, which are for a frequency of 500 Hz. By making some rather drastic simplifications in the theory of porous absorbers it can be shown that the thickness beyond which little improvement is obtained in absorption coefficient (the "sufficient thickness") is given *for a frequency of 500 Hz* by

$$L_{\text{sufficient}} = \frac{10}{\sqrt{r}}$$

where r is the flow resistance in M.K.S. units and L is in metres. Table 7.1 gives values for the flow resistance and sufficient thickness for a number of commonly used materials.

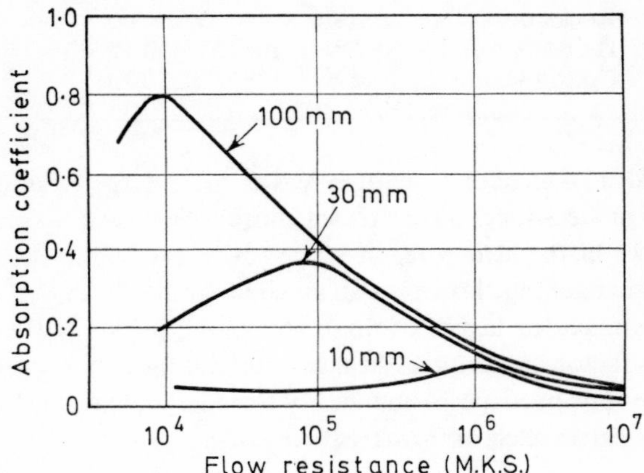

Figure 7.9. Optimum flow resistance at various thicknesses for frequency of 500 Hz.

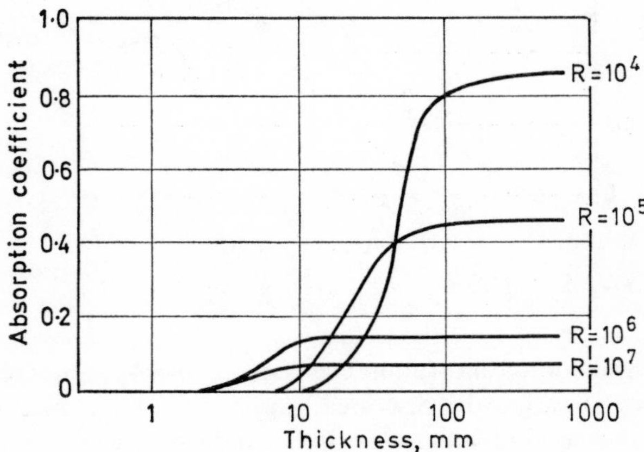

Figure 7.10. Sufficient thickness at various flow resistances for frequency of 500 Hz.

TABLE 7.1

Material	*r* (*M.K.S.*)	L *sufficient* (*m*)
Insulite	$1{\cdot}4 \times 10^7$	0·0085
Rockwool	$1{\cdot}28 \times 10^5$	0·09
Hair felt	$2{\cdot}9 \times 10^4$	0·19
Cotton wool	$1{\cdot}6 \times 10^3$	0·79

The effect of structure factor is less clear. A large structure factor tends to make sound waves travel more slowly and so emphasises resonances in the structure, produced by the standing wave set up by the rigid backing. There is also an effect similar to that of optimum flow resistance for layers of finite thickness; for an infinite layer a structure factor of one gives best results, but larger structure factors may give improved performance for smaller thicknesses. Structure factor has little effect at low frequencies.

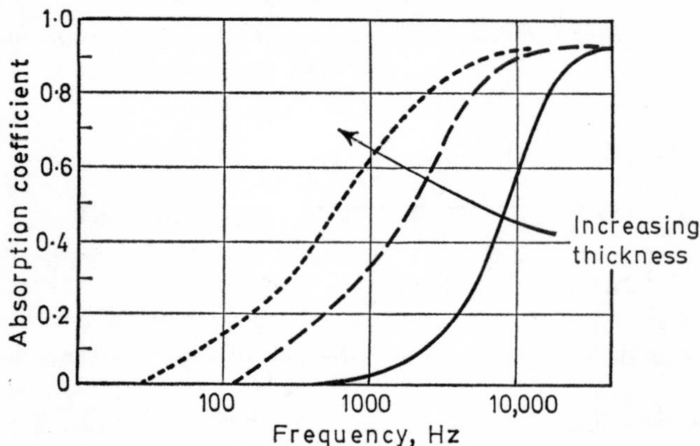

Figure 7.11. Effect of thickness on absorption coefficient of porous absorber.

The variation of absorption coefficient of porous absorbers with frequency is still to be discussed. Approximately, it may be said that an increase in frequency with the consequent decrease in wavelength makes a layer effectively thicker and decreases the characteristic impedance of the layer so that more incident sound energy gets in initially. These effects in combination lead to the kind of variation of absorption coefficient with frequency and thickness shown in Figure 7.11. In all cases for rigid porous absorbers the highest absorption coefficients are found at high frequencies. The performance at low frequencies will depend on the thickness of the layer and on the flow resistance of the material.

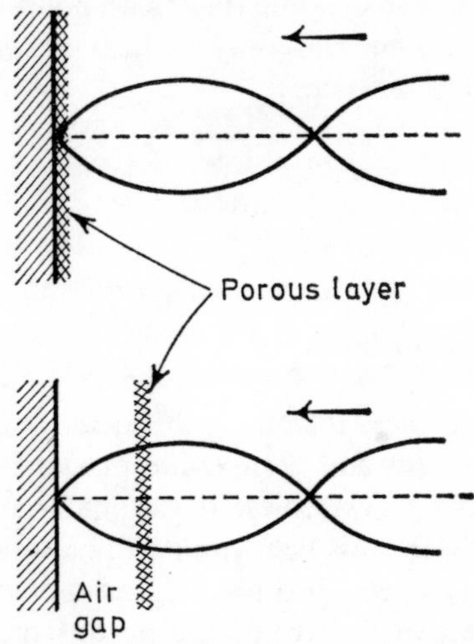

Figure 7.12. Effect of spacing porous layer away from wall.

One further point in connection with porous absorbers is that an increase in the mid-frequency performance of a layer can be obtained by placing it at some distance from the rigid backing wall. In this case its behaviour is almost equivalent to that of a complete layer of thickness equal to the sum of layer and air gap. This can be understood by reference to Figure 7.12, if it is remembered that maximum airflow in a standing wave occurs at a quarter wavelength distance from a reflecting wall. One disadvantage with this procedure is that a fall in absorption occurs at frequencies which produce standing wave patterns with a node at the absorptive layer, that is when the air gap is an integral number of half-wavelengths.

Finally it should now be apparent that the painting of a porous material will have a very deleterious effect on its absorptive performance. Anything which clogs the pores will increase the flow resistance at the front surface, and thus increase the fraction of sound energy which is reflected there. Tinting of the base material is permitted, of course, provided that this does not clog the pores. Friable materials may be protected with open screens (e.g. of expanded metal) provided they do not increase the flow resistance. A perforated panel less than 5 mm thick with holes b mm in diameter and separated by d mm will only reduce absorption power at frequencies above f kHz where

$$f = \frac{35b}{d^2}$$

7.8 Flexible Porous Surfaces

Flexible porous surfaces include such things as mineral and glass fibre mats, plastic foams and vermiculite plasters. They lie, in their behaviour, between two extremes. If the material is truly porous, with a low flow resistance its behaviour will be similar to that of the rigid porous materials described above, except that there will be an increase in the effective density of the porous medium due to the movement of the flexible frame through which the pores run. Elastic losses in this frame will also contribute to the absorption of the sound energy. This will obviously modify the requirements with regard to "sufficient thickness" or optimum flow resistance, although the extent to which this is the case will depend on how flexible the material is in relation to its flow resistance—mineral wool for example probably does not depart too far from a rigid porous material so that the theory of optimum flow resistance is probably not too far out. On the other hand, for many types of plastic foam the bubbles of gas generated in the manufacturing process remain intact after setting so that the pores fail to interconnect. In this case flow resistance becomes very high and absorption by motion of the frame becomes much more important. In the extreme the material becomes simply a flexible layer which yields under sound pressure without allowing penetration of its surface. Of course this sealed type of material (expanded polystyrene is an example) can be painted with a thin coat of paint without affecting its properties.

The variation of absorption coefficient with frequency for flexible porous materials is similar to that of rigid porous materials described above, except that standing wave effects are more pronounced, giving rise to fluctuations in the curve. The flexibility in the frame does tend to improve the absorption at low frequencies, though the amount of improvement will depend strongly on the thickness and flexibility of the layer. A typical curve of absorption coefficient versus frequency is shown in Figure 7.13.

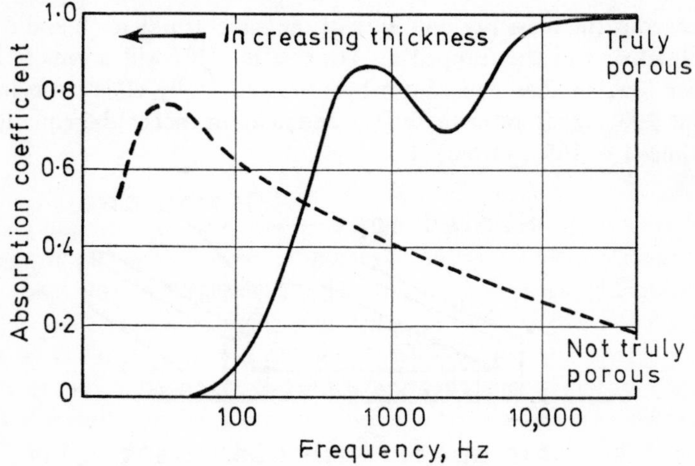

Figure 7.13. Absorption coefficient of a flexible porous absorber.

7.9 Impervious Flexible Surface

A sound wave incident on a flexible surface will set it into vibration. As it vibrates it works against the friction in itself, in its supports and in any space behind it. The vibrations excited in the system will be greatest, and hence the amount of energy absorbed will be greatest, when the frequency of the incoming wave matches the resonant frequency of the surface. At other frequencies the absorption will be poor.

The frequency of the panel, and hence the frequency of maximum absorption, will be determined by its own stiffness, the stiffness of its support, and its superficial mass. In a good many structural elements, such as windows, structural ceilings and floors the controlling stiffness is that of the surface itself. However, in many other cases air stiffness controls the resonance. In this case the panel bounces on a layer of air trapped behind the surface by a second, rigid surface. Then the resonant frequency can be determined from the expression

$$f = 60 \sqrt{\frac{1}{md}} \text{ Hz}$$

where m is the mass per unit area of the panel (in kg m^{-2}) and d is the thickness of the trapped air layer in m. This will normally be rather low, so that panel absorbers are only fully effective below about 200 Hz. If m is in lb ft^{-2} and d is in inches the equation becomes $f = 165 \sqrt{(1/md)}$ Hz.

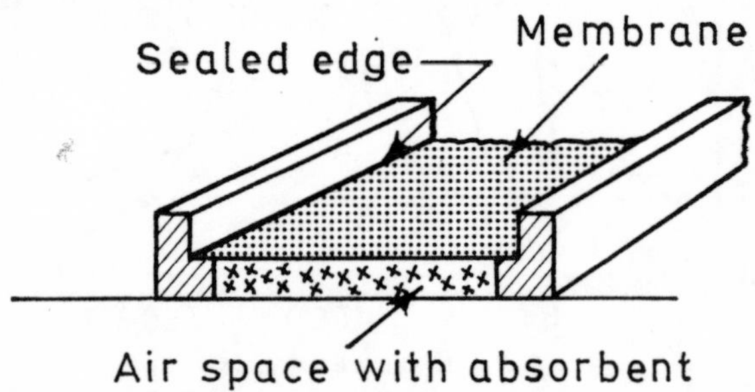

Figure 7.14. Membrane absorber.

The range over which the absorber acts (breadth of the resonance) will depend on the amount of damping (friction) against which it is working. A heavily damped system will act over a wider frequency range, but will be less responsive and hence less absorptive at its resonant frequency. Damping is provided by the supports and the internal friction of the layer, but extra friction can be provided by filling the trapped air space, if one exists, with a porous absorber. This type of structure thus complements porous absorbers very effectively, since it provides low frequency absorption without much high frequency absorption (*see* Figures 7.14 and 7.15). As already stated many conventional building structures exhibit this type of behaviour. A window, for example, of area $1{\cdot}25 \times 0{\cdot}85\ m^2$, and 3 mm thick exhibits the following absorption coefficients:

125 Hz	250 Hz	500 Hz	1 kHz	2 kHz	4 kHz
0·35	0·25	0·18	0·12	0·07	0·04

It is for this reason that structures such as churches, cellars and tunnels, that do not have freely vibrating panels, often show a marked lack of low frequency absorption not observed in buildings with large windows, wooden floors or panelling.

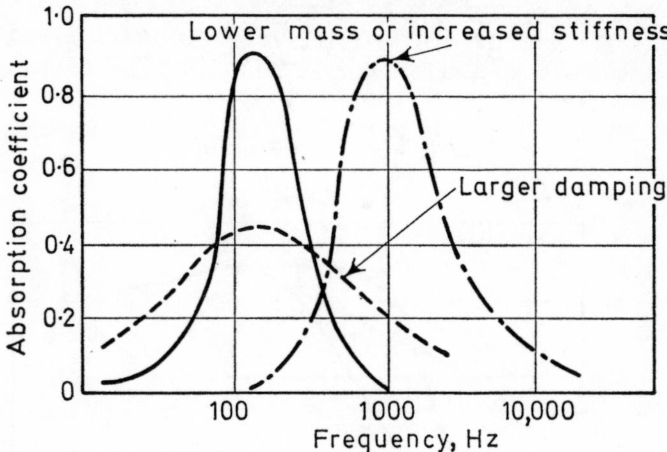

Figure 7.15. Absorption coefficient of typical membrane absorber.

Care must be taken if low frequency absorbers of small (that is small compared to wavelength) lateral dimensions are constructed, to ensure that the backing air space is well sealed. If air can escape to the side when the panel is pushed in, a lowering of the resonant frequency will result.

7.10 Acoustic Tile

The origin of the need for compromise in the selection of flow resistance, which is discussed in section 7.7, lay in the need to get sound into the material before attenuating it as rapidly as possible. A technique for avoiding the need for compromise, patented by Trader in 1922, consists of drilling holes down into the surface of a porous absorber of rather high flow resistance. This enables the sound to "penetrate" into the material where it may be rapidly

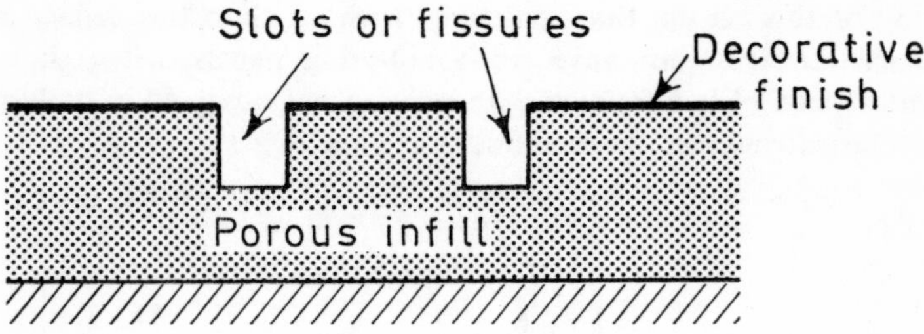

Figure 7.16. The acoustic tile.

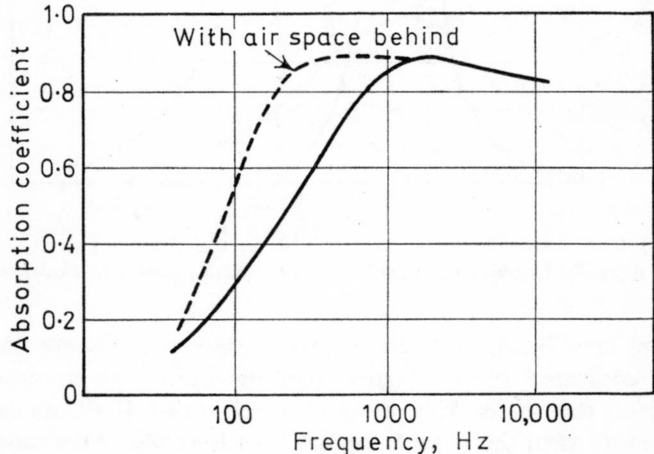

Figure 7.17. Absorption coefficient of typical acoustic tile.

absorbed. Since the expiration of the patent (originally held by the Celotex Corporation) many types of tile have come on to the market, the common factors being a penetrated surface and a high flow resistance. The shape of the surface cavities varies from the original cylindrical holes to include slits of various widths and even irregular cavities simulating travertine marble! The surface between the holes may be decorated.

The mechanism of absorption is rather complex and not very susceptible to analysis. From one point of view the material can be regarded as a low flow resistance porous absorber (the main holes) with a very large structure factor (the dead space between the holes being filled with the base material). It may also be regarded as a perforated plate resonator (*see* below) with the base material as infill. The design is still carried out to a more or less empirical "recipe". Certainly the improved performance at mid-frequencies seems to indicate that some kind of resonance is taking place in the artificial cavities (*see* Figures 7.16 and 7.17).

7.11 The Single Resonator

This type of absorber depends for its action on the flow of air in and out of a cavity which is connected to the room in which the sound is present through a restricted neck. Such an arrangement (rather like a narrow-necked bottle) again has a "resonant frequency" at which it sounds when stimulated, and to which it reacts strongly. Sound of the resonant frequency or close to it works against the friction of air in the neck, which "bounces" in and out by compressing and stretching the air in the cavity. It is usual to increase this resistance artificially by the addition of porous material to the neck, or by loosely filling the cavity with fibrous material to impede the movement of air. This is done for two reasons: first, unless the natural viscosity of air in the neck is sufficiently large the system acts as a store of energy instead of dissipating it, so that when the source of sound is cut off the energy is re-radiated to the room, prolonging the reverberation (*see* below). Secondly, the degree to which the effect of the resonator spreads to frequencies adjacent to the resonant frequency is once more a function of the amount of resistance as with the membrane absorber. This may be provided in the neck, or by filling the reservoir loosely with some fibrous material. By increasing the resistance the absorption can be spread over a wider frequency range but this again reduces its effect at the resonant frequency just as for the panel absorber. The resonant frequency depends in a rather complicated but well documented way on the dimensions of the cavity and the neck. In addition to the single resonator there is an alternative form which consists of a slit opening into a long cavity. This acts in the same way, but on a "per unit length" basis. The most important thing to notice about the resonator absorber however is that its effective absorbing area

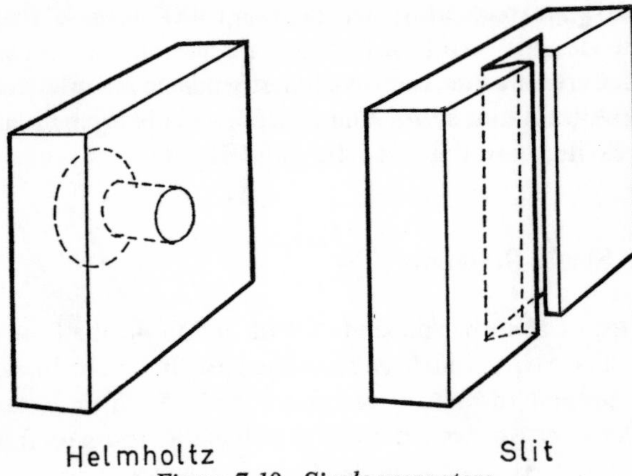

Figure 7.18. Single resonators.

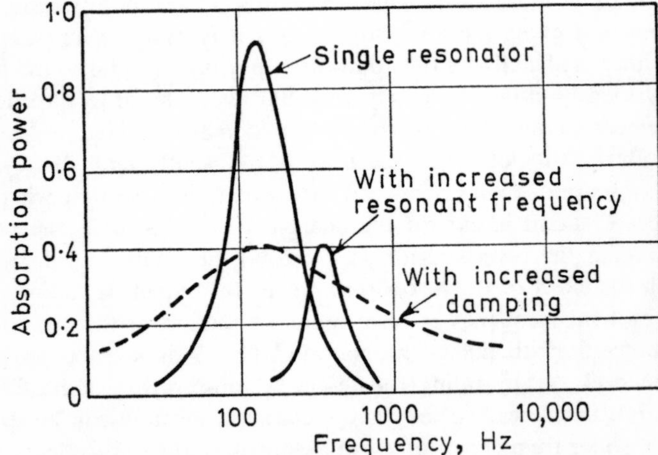

Figure 7.19. Absorption power of typical single resonators.

is very much larger than the area of the neck—it is in fact proportional to the square of the wavelength. For this reason resonators are used only for low frequencies where the wavelength is sufficiently long to make them an economic proposition. Typical designs and frequency/absorption curves are given in Figures 7.18 and 7.19.

7.12 Perforated Plate Multi-Resonators (*see* Figures 7.20 and 7.21).

At higher frequencies the provision of separate resonators becomes uneconomic because of the dependence of absorption power on wavelength. However, it is found that the adjacent walls between

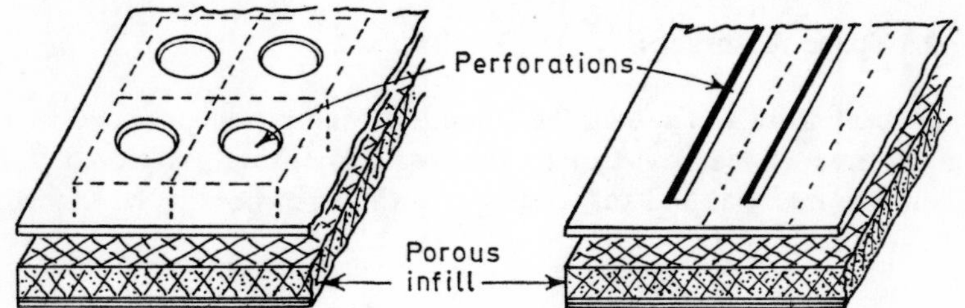

Figure 7.20. Multiple resonators.

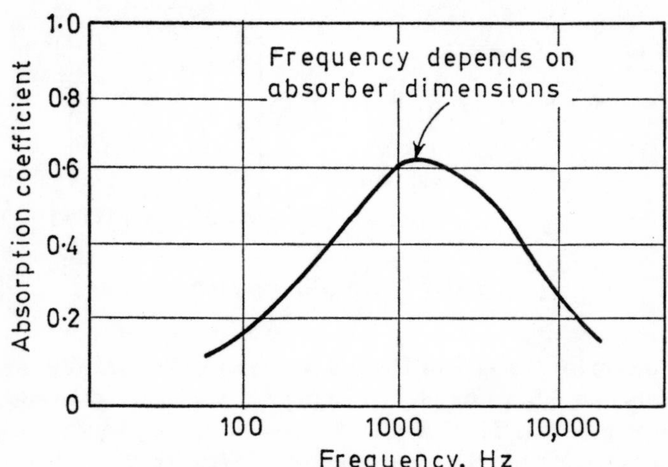

Figure 7.21. Absorption coefficient of multiple resonator.

small resonators can be omitted, the sound waves automatically dividing the continuous space behind a perforated panel into a series of "cavities" to match the "necks" of the separate perforations. Once again it is necessary to "spread" the absorption by providing resistance, usually in the form of fibrous material in the space behind the panel. The perforations may be in the form of holes or slits; the design parameters are well documented. Care should be taken to differentiate between this type of absorbent and the normal porous absorbent with protective open work cover. In the case of perforated plate resonators, the open area should be less than 15%, while for "open" covers the area should be above 30%.

7.13 Space Absorbers

Recent work has shown that the absorbing power provided by a given area of material can be increased by forming a closed box from the material and suspending it within the room to be treated.

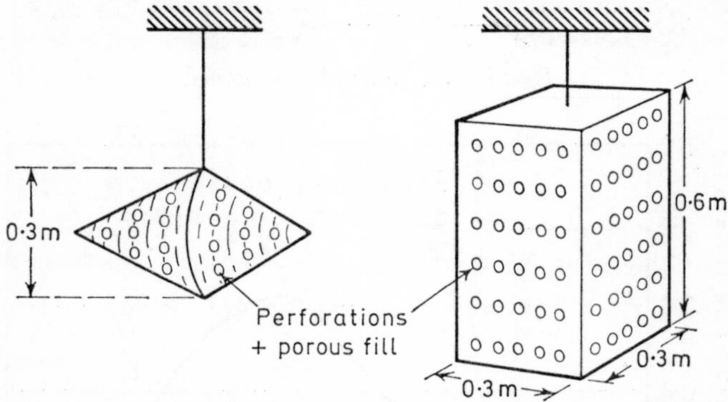

Figure 7.22. Space absorbers.

The theory of the system has not yet been developed fully but it appears that diffraction plays a large part in increasing the effective absorbing power. There is also a resonance effect, which causes a greater increase to take place at lower frequencies.

They suffer from the defect that they are rather obtrusive visually, and so are probably best restricted to workshops and lower-grade offices. A typical design and absorption curve appears in Figures 7.22 and 7.23.

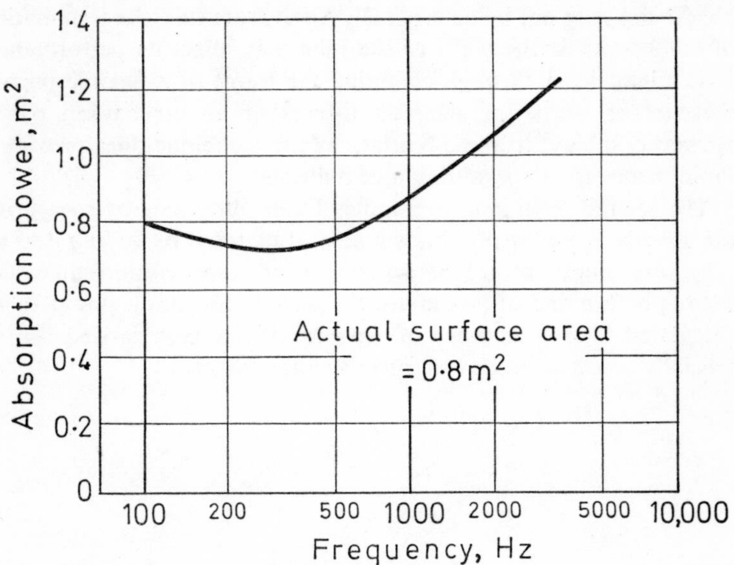

Figure 7.23. Absorption powers of typical space absorber.

7.14 Methods of Measurement of Absorption Coefficient and Absorptive Power

There are basically two ways of measuring absorptive properties. The first, and simplest to perform, consists of placing a small sample of the material under test at the end of a tube. A standing wave is then produced by sending a sound wave down the tube and reflecting it from the sample. The relative sizes of the "nodes" and "antinodes" in the pattern give a measure of the absorptive properties of the sample, for normal incidence.

Unfortunately this method only gives data for sound reflected "head-on" from the sample. In practice, in a room, sound may strike the material from any angle, and the absorptive properties usually vary with angle of reflection. There is a simple theory which can extend tube results to more realistic conditions, but it is not applicable to a good many real materials. Another objection to the tube method is that the absorption due to whole objects (e.g. people or space absorbers) may not be determined. Apart from this, the termination of the sample at the walls of the tube may affect its performance. This might occur through clamping the frame of a flexible porous material or increasing the self-stiffness of an air backed panel. However, subject to these limitations the technique does provide a rapid means for the evaluation of materials.

The second technique overcomes these objections by measuring the absorbing power of a known area of material by adding it to an otherwise empty room. Measurement of reverberation time (*see* below) before and after enables the added absorption power to be calculated. For a number of reasons this is best carried out in specially constructed rooms (reverberation rooms).

CHAPTER 8

Room Acoustics

8.1 Introduction

In this chapter the behaviour of sound in rooms is first considered in general terms. A calculation is used to illustrate the application of these ideas to a specific problem. Finally, the requirements for specific types of auditoria and other halls are discussed.

8.2 Standing Waves in Rooms

If sound is supplied to a long narrow pipe, closed at both ends and the frequency of the sound is varied, the pipe will respond or "resonate" at certain frequencies, whenever the half-wavelength of the sound "fits" a whole number of times into the length of the tube. These are also the frequencies which are produced when the tube is "excited" for example by blowing down it so that it acts like an organ pipe. The pressure system set up inside the tube when it is resonating is referred to as a standing wave. Similar standing waves can also be set up between the walls of a room, not only between opposite, parallel walls, but also by successive reflection from end walls, floor and ceiling in a diagonal standing wave, or even in a path taking in all six surfaces. For simple rectangular rooms it is possible to calculate the frequencies of these standing waves. It is found that at low frequencies the resonant frequencies are rather widely spaced so that it is easy to pick out the tones that set the room into strong vibration. At higher frequencies the wavelength of the sound becomes so short that it is possible to fit all frequencies in as a standing wave in some way or other, so that the resonances become very much closer spaced and less easy to pick out. The frequency at which this blending of the room response occurs depends very much on the size of the room. In a bathroom, for example, it is very easy to hum the first few resonant notes, but in larger rooms this is not possible. The simple mathematical analysis which applies to rectangular rooms is not available for more complex room shapes, but although they cannot be predicted easily, resonant frequencies still exist. Once again, the closeness of the spacing of these frequencies is a function of the size of the room. A sound source which radiates at one of these frequencies will tend to couple very strongly and consequently become more efficient so that a greater level is produced.

8.3 Reverberation in Rooms

If a source of sound were to be started up in a room with perfectly reflecting walls and with no air absorption the amount of sound energy in the room would build up continuously, the sound getting louder and louder without limit. Fortunately all surfaces absorb some fraction of the sound energy incident on them. Since the total amount of sound energy absorbed per second depends directly on the sound energy incident on the walls per second; the rate of absorption depends on the intensity of sound in the room. So after the sound source starts up the intensity of sound in the room increases until the rate of absorption of energy balances the rate of production, and a steady state is achieved. The intensity of the sound in the room at this steady state is proportional to the power of the source and inversely proportional to the total amount of sound absorbing material at the walls.

The sound-absorbing power is calculated by adding together the product of area and absorption coefficient for all the surfaces in the room together with an equivalent area for the air absorption. Doubling the absorbing power by the addition of, for example, acoustic tiles, will halve the intensity in the room, but this is only a reduction of intensity level of 3 dB. Obviously it will not be possible to achieve even this if the room is already reasonably well supplied with absorbent in the form of carpets and well-upholstered furniture.

When the sound source has just been switched on the rate at which the intensity in the room builds up will depend on how close it is to its final value, for as the intensity approaches the stable condition there is less and less energy left over after reflection from the walls to continue "topping up" the reverberant sound. This gives rise to an "exponential" growth in sound level. The reverse process occurs when the sound source is cut off. Initially the sound is absorbed rapidly by the walls of the room but as the remaining level falls, the rate of absorption becomes less so that there is an exponential decay of sound. If this exponential growth and decay of sound is regarded on a logarithmic basis the growth is seen as an almost instantaneous rise while the decay becomes a "linear" one with the sound pressure level falling off proportionally with time. The sound field built up in the room by the source is referred to as the "reverberant" field, and the persistence of this sound after the cut-off of the source is termed reverberation. It is found that the time that the reverberant field takes to build up and die away is proportional to the volume of the room and inversely proportional to the total absorption (area × absorption coefficient). Just as absorption coefficient can vary with frequency, so can reverberation time, i.e. a full specification of reverberation must cover variation with frequency. The reverberation time is precisely defined as the time taken for the sound level to fall by 60 dB after the cut-off of the source. In this case

$$\text{reverberation time} = \frac{0{\cdot}05 \times \text{volume of room in cu. ft}}{\text{total absorption power in sq. ft}}$$

or

$$= \frac{0{\cdot}16 \times \text{volume of room in cu. m}}{\text{total absorption power in sq. m}}$$

This is often referred to as the Sabine equation.

The importance of reverberation time is twofold. In the first place it acts as a measure of the "noise sensitivity" of an enclosure. For a given volume, a long reverberation time corresponds to small total absorption power. But the total intensity of the reverberant sound produced while a source is operating continuously is inversely proportional to the total absorption. Halving the reverberation time corresponds to a doubling of the absorption power, and hence to reducing the reverberant sound level by 3 dB.

The second reason for the importance of reverberation time lies in its use as an acoustic criterion. This will be discussed in more detail later.

It should be made clear at this stage that reverberation is not quite as simple as the foregoing would indicate. The simple treatment of a room as a "leaky reservoir" of acoustic energy is complicated by the standing wave effects described at the outset. The total energy in the room at any time is described efficiently by the simple analysis, but when the sound energy is considered at one particular point in the room, the effect of the gradual decay of the various standing waves which have pressure maxima at this point is to produce a violent fluctuation in intensity. This is a phenomenon akin to the "beats" heard when two notes of almost equal frequency are sounded together. For this reason the decay of sound level with time at one point in a room is not in fact linear but fluctuates about the linear decay. The procedure in this case is to draw the best straight line through a trace of the decay, taking the average slope from a large number of trials. Another technique which aids this averaging process is to excite the room with a "band" of frequencies simultaneously, using either a rapidly warbled pure tone or bands of random noise. In some circumstances it may be permissible to use broad band noise (e.g. pistol shot) to excite the room, but to filter the received sound to examine the decay at particular frequencies.

8.4 Room Radius

Close to the source of sound in a room, the sound pressure direct from the source will have a larger level than the reverberant sound. This is called the direct sound field. The distance from the source at which the direct sound field falls (because the energy is spreading out spherically) to a level equal to the reverberant sound field is termed the room radius, and is given by

$$r = \sqrt{\frac{A}{16\pi}}$$

where A is the total absorption in the room.

Reduction of the sound level at a distance *less* than the room radius from a source of noise cannot be achieved by absorbent treatment of the walls, since the direct sound predominates.

Another parameter that can be measured in a room is its spatial diffusion. This is a term which describes the degree to which a uniform sound level exists at all points in the room while the source is sounding. If the sound field remains diffuse during the decay process it is sufficient to measure reverberation time at only one place, since this is then typical of the decay of energy throughout the whole room. A related problem concerns directional diffusion; this is concerned with the degree of equality between the sound energy arriving at a point from different directions. If directional diffusion is poor the decay of reverberant sound may be complicated by different rates of absorption of sound waves travelling in different directions. This leads to "multiple-sloped" decay curves.

8.5 Other Acoustic Defects

In addition to the above considerations there are a number of peculiarities in the acoustics of rooms which arise because of their shape.

Echoes.—An echo arises whenever a strong reflection of an original impulse arrives at the ear after an interval of greater than 0·05 seconds. For intervals between original and reflected sound arrivals less than this, the reflected sound is not recognised as a separate echo but adds to the apparent strength of the original sound. Strong reflections arise, particularly from concave surfaces, which are able to focus reflected sound to a particular point. The time taken for a particular reflection to arrive at a point in the auditorium can be determined from ray diagrams. The difference in time between the direct and reflected sound is obtained by dividing the difference between their path lengths in metres by 340 (or feet by 1100).

Whispering galleries.—These arise when a source is placed very close to a long curved surface. The sound waves starting tangential to the wall "creep" by progressive reflection with very little attenuation to points around the curved surface.

Dead spots.—Excessive concentration of sound due to curved surfaces must result in a comparative deficiency elsewhere. Positions in a room far removed from reflecting surfaces with sound arriving only after "grazing" over an audience for some distance, are also likely to constitute dead spots.

Flutter.—This arises when source and listener are both between a pair of hard parallel surfaces and other surfaces in the vicinity are fairly absorbent. It can be avoided by providing absorbent on the parallel walls or by breaking up the surfaces with ornamentation or large scale moulding.

8.6 Example of the Calculation of Reverberation Time

In many cases the simple Sabine equation given above gives a sufficiently accurate value of reverberation time. The following example illustrates its application to the acoustic treatment of a swimming pool.

A rectangular swimming pool is of length 60 m (200 ft), width 18 m (60 ft) and height 9 m (30 ft). One long wall is glazed from floor to ceiling. The end walls are of rendered concrete. The other wall has changing lockers with wooden doors to a height of 2·6 m (8 ft) at floor level, the depth of the lockers being 2 m (6 ft). Above this is a spectators gallery with a wooden floor and two rows of benches. The wall above this to ceiling height is again of rendered concrete. The ceiling itself is of structural concrete, sloping to a 1 m (3 ft) rise at the centre. The pool is 50 m $\times$ 12 m (150 ft $\times$ 40 ft). The floor around the pool is of rubber tiling on a sub-floor of concrete.

Using the absorption coefficients given opposite calculate the reverberation times, and the room radius and the reverberant intensity produced by a source of 1 mW acoustic power with a frequency of 500 Hz. Discuss the treatment of the room.

8.7 Auditoria for Speech

Quite apart from questions of the "acoustic ambience" of a room for speech, the prime requirement is that what is said should be

Frequency	125 Hz	250 Hz	500 Hz	1 kHz	2 kHz	4 kHz
Item	*Absorption coefficients*					
Plate glass as window	0·18	0·06	0·04	0·03	0·02	0·02
Wooden floor on air space	0·15	0·11	0·10	0·07	0·06	0·07
Doors on air space	0·30	0·20	0·10	0·07	0·06	0·07
Smooth concrete	0·02	0·02	0·02	0·03	0·04	0·05
Water	0·008	0·008	0·013	0·015	0·020	0·025
Rubber floor tiles	—	0·05	0·05	0·10	0·05	—
Air absorption $\equiv$ m² per 100 m³ of vol.	—	—	—	—	0·75	2·3
$\equiv$ ft² per 1000 ft³	—	—	—	—	2·3	7·2
Benches, m² per metre run	0·02	0·03	0·03	0·06	0·06	0·05
Benches, ft² per ft run	0·07	0·09	0·10	0·20	0·20	0·15

Solution: First calculate (area × absorption coefficient) for all the surfaces

Item	*Area, m²*	*Absorption power* 125	250	500	1000	2000	4000
Windows	540 m²	97·2	32·4	21·6	16·2	10·8	10·8
Wooden (gallery) floor	120 m²	18·0	13·2	12·0	8·4	7·2	8·4
Locker doors	156 m²	46·8	31·2	15·6	10·9	9·4	10·9
Concrete walls and ceiling	1806 m²	36·1	36·1	36·1	54·3	72·2	90·4
Water	600 m²	4·8	4·8	7·8	9·0	12·0	15·0
Tiled floor	360 m²	—	18·0	18·0	36·0	18·0	—
Air absorption	Vol. 10,260 m³	—	—	—	—	76·9	236·0
Benches	120 m length	2·4	3·6	3·6	7·2	7·2	6·0
Total	m²	205·3	139·3	114·7	142·0	213·7	377·5
T_{60} (sec)		8·0	11·8	14·3	11·6	7·7	4·3

$$I_{rev} = \frac{4W}{A} = 3{\cdot}5 \times 10^{-5} \text{ watts m}^{-2}$$

$$\text{room radius} = \sqrt{\frac{A}{16\pi}} = 1{\cdot}7 \text{ m}$$

clearly and easily understood. The degree of clarity at various points in an auditorium can be measured by an articulation test. This consists in the reading of lists of unconnected single-syllable words from the speaker's platform; observers at various parts of the auditorium write down what they hear, or guess at the words they miss. An analysis of the percentages of vowels and consonants correctly heard is then carried out to give the "percentage syllable articulation" (P.S.A.). Obviously the intelligibility of connected speech is not impaired if some syllables are not understood, because of the deductive ability of the listener. In fact, in perfect listening conditions a P.S.A. of 95% is normally the maximum obtained owing to unavoidable errors. However, with P.S.A.'s of less than 75% some concentration on the part of the listener is necessary to understand what is said, while below 65%, conditions are definitely unsatisfactory.

It is generally reckoned that four factors contribute to articulation at any point in the room. The first is the background noise level which "masks" the required sound. This should be kept below 30 dB(A) if speech is unaided by amplification systems. The second factor is the level of the speech above the threshold of hearing. This will depend (assuming no artificial amplification is provided) on the distance from the speaker, the volume of the hall (inversely, through the reverberant sound field) and the nature of the speaker's surroundings (whether they are highly reflecting or not). The third factor is the reverberation time; if this is very short there will be insufficient reverberant energy to maintain the level of speech. On the other hand, if it becomes too long the sound of successive syllables overlaps, and the resulting reverberation "masks" the speech. The last factor is the shape of the room, though if this has been designed to avoid echoes and dead spots and to give good vision of the speaker from all parts of the auditorium, it should have no effect upon the articulation.

Returning to the question of the masking of words by the reverberation of previous words, the difference between "early reflected sound" and "reverberation" must be discussed further. It is found that if a reflection or "echo" of an original sound arrives at the listener's ear less than about 0·05 sec after the original sound, it is not perceived as a separate echo but simply serves to increase the apparent loudness of the original sound (this is known as the Haas effect). Reflections that arrive later than this constitute the "reverberation", which serves only to mask the next syllable, or if late enough and strong enough, constitute a separate "echo". For this reason reflections can be divided into helpful and harmful categories. Those which arise from surfaces close to the speaker or listener increase intelligibility because they boost the loudness of the direct sound, while general, prolonged reverberation tends to decrease intelligibility.

To some extent, a compromise is required between loudness (low absorption) and short reverberation time (large absorption). The optimum value varies with volume (*see* Figure 8.1); this value should apply throughout the frequency range from 125 to 4000 Hz. If reverberation time increases towards low frequencies there may be a tendency towards "booming", while if high frequencies have long reverberation times a harsh, dry quality may result.

Before leaving the discussion of speech auditoria some mention should be made of electronic reinforcement systems. As a rough guide, it can be said that reinforcement will be required in rooms which do not have specially shaped walls and ceilings, (*a*) for speaker–listener distances greater than 12 m (40 ft) or (*b*) for volumes greater than 400 m^3 (15,000 ft^3). If, however, the speaker is aided by correctly positioned reflectors these figures increase to 24 m (80 ft) and 2000 m^3 (75,000 ft^3) respectively. Calculation of the required number of loudspeakers and power of the system is complex, taking account of background noise and the volume and reverberation time of the room. If the speech is not to sound unnatural it is important that the amplified sound should arrive at the listener just after the sound which has travelled direct from the speaker. The Haas effect then ensures that the apparent source remains at the speaker. This is most easily achieved by placing loudspeaker outlets above and behind the speaker. However, this is ruled out in highly reverberant conditions, since a small number of high power outlets increase both reverberant and direct sound, doing nothing to improve articulation. In this case many small outlets may be placed around the auditorium, each fed with an electronically delayed signal so that the sound from them arrives after the direct sound from the speaker, even though the outlets are closer. Furthermore,

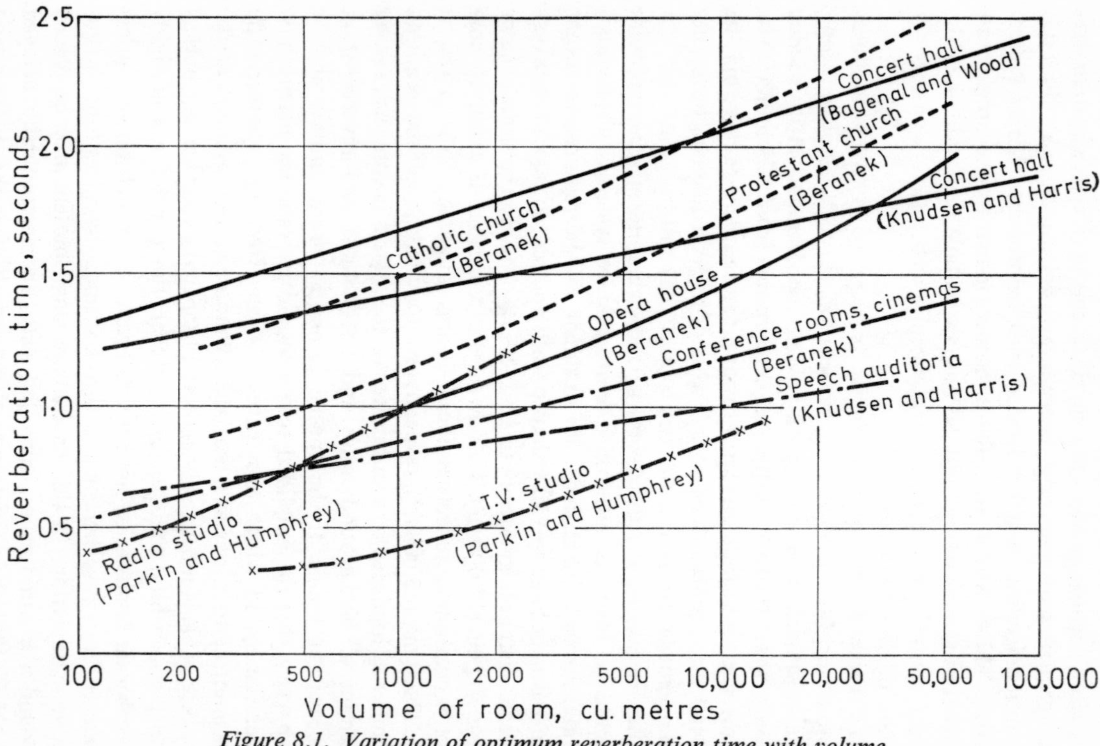

Figure 8.1. Variation of optimum reverberation time with volume.

by using assemblies such as column loudspeakers which beam sound on to the audience, the reinforcing signal can be absorbed (the audience is a fairly efficient sound absorbing surface) and so prevented from adding to the reverberant level. In this way the ratio of direct to reverberant sound can be increased. This technique is particularly useful for enclosures with unavoidably long reverberation times, such as churches and sports halls.

8.8 Concert Halls

The subjective acoustic properties of concert halls are unfortunately not sufficiently well defined to enable exact prediction. The main variable is the reverberation time, but even when this lies in the recommended region (*see* Figure 8.1) the acoustics may not be accepted as fully satisfactory. It is generally agreed that the reverberation time should increase at low frequencies to provide "warmth" of tone, and that there should be some early reflections in order to give "presence" or "attack". Other requirements which are often spoken of are "ensemble" or "blend" and "clarity of tone". These are probably related in some way to the ratio of early reflected to reverberant sound, though the precise relations have not been satisfactorily established.

Avoiding these rather uncertain aspects, the known factors can be stated as follows: the reverberation time at mid-frequencies should be within the range shown in Figure 8.1, but there should be some increase at low frequencies. In addition, in large halls it may be necessary to provide a reflector above the orchestra platform in order to increase the proportion of early reflected to reverberant sound.

8.9 Opera Houses

An opera house must to some extent represent a compromise between articulation and musical sound. Arrangements must also be made to adjust the balance between singer and orchestra in favour of the singers. This is achieved by careful design of the orchestra pit. The reverberation times recommended for opera houses are shown in Figure 8.1.

8.10 Legitimate Theatre

Here articulation is the only requirement, so that optimum reverberation can be selected as for speech auditoria. However, the situation is usually complicated by the highly absorbing surroundings of stage sets and the loss of sound into the "flies". All the considerations with regard to echoes, dead spots, etc. apply with especial force to the live theatre.

8.11 Cinemas

The acoustical properties of the cinema auditorium must not modify the sound as recorded on the film. For this reason reverberation times are usually quite short. Modern practice is tending towards reverberation times of less than 1 sec, even in the largest theatres. Speech levels are much higher than in the live theatre so that higher background noise levels can be tolerated. It is, however, very important to isolate the projection booth, providing its interior with lots of absorption, and using double glazing in the projection windows.

8.12 Churches

The primary consideration in regard to churches must always be insulation against outside noises. In their earliest environment there was very little outside noise to deal with. Nevertheless, with traditional methods of construction of heavy walls, small windows, and tortuous entrance very good sound insulation was usually provided. In many modern churches, however, noisy environment and unconventional architectural design combine to give noisy interiors. For this reason a new design must be examined closely in this respect.

Almost as important is the control of reverberation. Conventional styles in churches of different denominations lead to a wide range of reverberation times, Roman Catholic churches having longer reverberation times than Protestant buildings. This arises from the distinct architectural styles, and reflects the relative importance of articulation, and organ and choral music. Optimum values of reverberation time for various volumes are shown in Figure 8.1.

In very large churches and cathedrals control of reverberation time becomes difficult because of the enormous volume and the problem of finding architecturally acceptable absorbent materials. Resonator absorbers have been used, particularly in Europe.

8.13 Gymnasia and Swimming Pools

These are only mentioned in so far as they can become a serious source of annoyance if sited close to residential areas, and in that it is sometimes necessary to provide for reasonable articulation if public competitions take place in them.

The control of reverberation is again very difficult, particularly in swimming pools where materials must be able to withstand the very high humidity. Vermiculite plasters and sprayed asbestos are fairly successful and resonator absorbers have also been applied for low-frequency control. Reverberation times should be not more than twice that for auditoria of comparable volume if articulation is to be higher than 65%, though public address systems can help.

8.14 Multi-purpose Halls

The multi-purpose hall presents a particular difficulty in that the main variable, reverberation time, has normally to be fixed for one purpose only. It is recommended that some sort of percentage rating be given for each of the possible uses of the hall, and a reverberation time chosen accordingly. In extreme cases it may be necessary to use surfaces of variable absorption or electronic devices for producing artificial reverberation.

8.15 Radio and Television Studios

Apart from studios for the performance of music, reverberation time is normally kept as short as possible, first to maintain high articulation (there is no problem of sound level) and secondly to reduce the level due to any outside noise which penetrates into the studio. This is the main consideration (especially for radio studios), but falls outside the scope of this chapter. Almost any highly absorbing material can be applied to the walls, since appearance is unimportant.

8.16 Landscaped Offices

Noise problems in normal offices have been dealt with elsewhere, but the recent trend to large open offices with acoustic ceilings and thick carpets introduces some novel problems in room acoustics. The propagation of sound in these offices is by direct transmission (giving inverse square law attenuation with distance) and by reflection between floor and ceiling. The walls are usually too far away to affect the sound field near the centre of the offices. At quite small distances from a noise source a sound wave is reflected from floor or ceiling at a grazing angle of incidence. The absorption coefficient of a surface is strongly dependent on angle of incidence, falling off sharply as the wave approaches grazing incidence. This is more marked at higher frequencies. Thus, although almost free field conditions may exist close to the source, at greater distances something like a $1/r$ dependence may set in because of the "images" of the source in floor and ceiling. This might be corrected by developing ceiling absorbers, or absorbing structures, with larger grazing-angle absorption coefficients. At still greater distances the level falls more rapidly again, though research needs to be carried out to establish controlling factors. Work is also required to establish permissible dimensions for rooms which are to exhibit wall-independence of the sound field.

CHAPTER 9

Criteria

9.1 Introduction

The establishment of criteria is basically an attempt to relate the physical measurements of sound to the human perception of sound. Existing criteria are conveniently divided into two categories:

(1) Those to protect individuals from noises which are potentially harmful.

(2) Those to protect individuals from noises which intrude and annoy.

Whatever the circumstances, it is always important to keep the acceptable noise environment in mind, because insulation or noise reduction costs money. The noise is most easily specified by a single dB(A) value which is recommended British practice, but this has the disadvantage that frequency dependent effects of partitions, silencers, absorbers, etc. cannot be taken into account. An alternative method is to specify the noise in nine octave bands centred on frequencies 31·5, 63, 125 to 8000 Hz. This is usually done graphically as, for example, in Figure 9.1 where a set of noise criteria (NC) curves are shown. The expected, or measured, noise is compared with the curves and the lowest curve which is nowhere exceeded by the noise gives the NC number.

In the case of dwellings, experience has shown that particular forms of construction, e.g. a brick wall, do provide an acceptable living environment. Consequently, building regulations suggest that constructions should give an insulation conforming to grading curves based on the insulation obtainable from traditional construction. These normally cover the frequency range 100 to 3150 Hz. The insulation of partitions for use in offices, hospitals, etc. has often been specified in terms of its average sound reduction index but occasionally a grading curve has to be met and it seems probable that grading curves will, in the future, become more common.

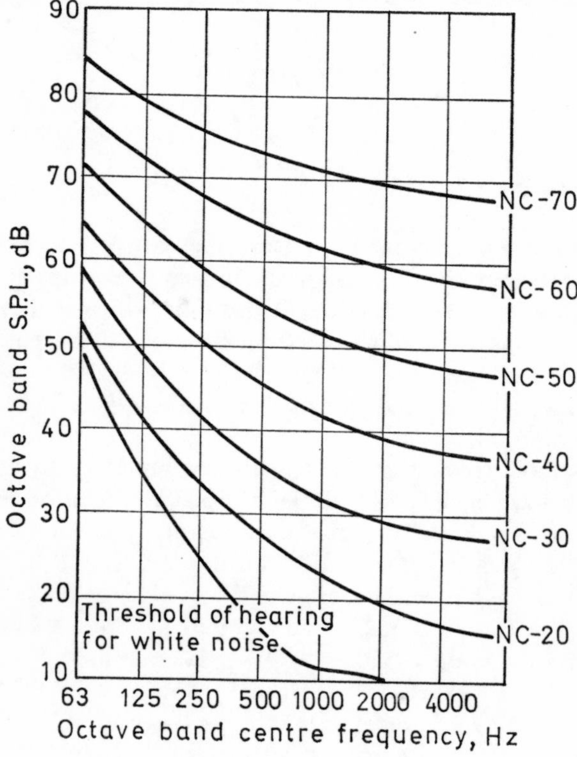

Figure 9.1. NC curves.

9.2 Noise which is Hazardous

Although exposure to loud noises may cause increases in blood pressure and heart rate, and in exceptional cases can affect the sense of balance, there is no evidence to suggest that any permanent damage to bodily function occurs. The exception to this generalization is the ear. Noises so loud that they cause immediate damage to the ear are not normally met with. However, a gradual deterioration, often not apparent to the individual until it is too late, is more likely to occur. The probability of permanent damage occurring appears to be a function of exposure time as well as noise level and tentative dB(A) levels which should not be exceeded are given in Table 9.1.

TABLE 9.1

SUGGESTED MAXIMUM EXPOSURE TIMES FOR AVOIDING DAMAGE TO HEARING IN A NOISY ENVIRONMENT

dB(A)	*Permissible exposure time in min/day*
90	500
95	140
100	50
105	30
110	17
115	10

For exposure times greater than 1 hour a good estimate of the maximum recommended exposure can be found on the basis that a doubling of energy (i.e. an addition of 3 dB) could be compensated by halving the duration of the exposure.

If the noise has a strong single frequency content or is very impulsive in nature it may well be more damaging than its dB(A) level would suggest, but as yet little is known about such niceties.

9.3 Noises which Intrude

9.3.1 *Rooms Other than Dwellings*

One of the earlier attempts to specify maximum acceptable background noise levels within rooms introduced the term Speech Interference Level (S.I.L.). The S.I.L. is defined as the average background noise level of the three octave bands 600 to 4800 Hz (American non-standard frequencies) at which speech communication can just take place.

Table 9.2 shows the S.I.L.'s in dB for a male voice. The levels should be reduced by 5 dB for a female voice.

TABLE 9.2

SPEECH INTERFERENCE LEVELS IN dB

Distance between speaker and listener		*Voice level*			
metres	*feet*	*Normal*	*Raised*	*Very loud*	*Shouting*
0·15	0·5	71	77	83	89
0·3	1	65	71	77	83
0·6	2	59	65	71	77
0·9	3	55	61	67	73
1·2	4	53	59	65	71
1·5	5	51	57	63	69
1·8	6	49	55	61	67
3·7	12	43	49	55	61

The S.I.L. is not always reliable because it only takes into account the noise in three octave bands and, although this is the most important range, low-frequency noise can be quite intrusive. It is therefore safer to take the complete frequency range into account by using, for example, the noise criteria curves of Figure 9.1. These NC numbers are in fact numerically equal to S.I.L.'s but they put rather stringent limits on the amount of low-frequency noise permitted for a given S.I.L.

The modern tendency is to specify an acceptable NC number (or some equivalent criterion) for a particular type of room. Table 9.3 lists some recommended noise criteria which should be acceptable.

9.3.2 *Dwellings*

The Wilson Committee report on noise gives some specific

guidance in this instance. Based partly on people's reaction to internal noise in and around London and partly on people's reaction to domestic noise from neighbouring dwellings, it is tentatively suggested that the noise levels shown in Table 9.4 should not be exceeded in living rooms and bedrooms for more than 10% of the time.

TABLE 9.3

RECOMMENDED BACKGROUND NOISE CRITERIA FOR DIFFERENT TYPES OF ROOM

Type of room	*NC number*
Hospital, theatre, church, cinema, concert hall, small office reading room, conference room, lecture room	20–30
Larger office, business store, department store, meeting room, quiet restaurant	30–40
Larger restaurant, secretarial office (with typewriter), gymnasium	40–50
Larger typing hall	50–60
Workshops	60–70

TABLE 9.4

RECOMMENDED NOISE LEVELS IN DWELLINGS WHICH SHOULD NOT BE EXCEEDED FOR MORE THAN 10% OF THE TIME

Situation	*Day*	*Night*
Country areas	40 dB(A)	30 dB(A)
Suburban areas, away from main traffic routes	45 dB(A)	35 dB(A)
Busy urban areas	50 dB(A)	35 dB(A)

9.3.3. *Insulation between Dwellings*

Figure 9.2 shows the recommended values of airborne sound insulation (United Kingdom) between houses. The measured insulation (corrected to a reverberation time of 0 5 sec, which is appropriate to the average living room) should come above the grading

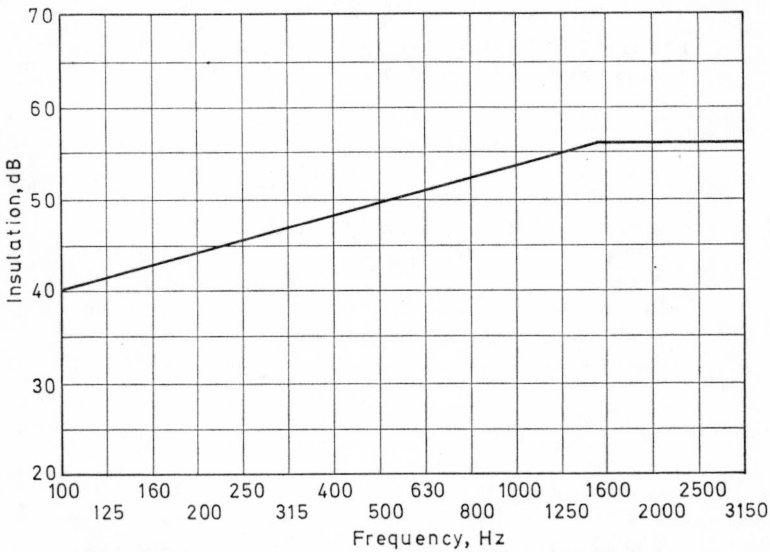

Figure 9.2. Recommended insulation between houses (U.K.).

line although a total unfavourable deviation of up to 23 dB is normally permitted.

It is thought that tenants of flats are a little more tolerant towards noise coming from neighbouring flats because it is only one of a number of minor disadvantages which they suffer. Figure 9.3 shows two grades of airborne sound insulation between flats and these apply to party floors as well as to party walls. Grade I represents the insulation which should be aimed at and, if achieved, noise from neighbours causes only minor disturbance to most tenants. Grade II insulation satisfies about 50% of tenants but the other 50% consider that noise from neighbours is the worst single factor about living in flats. Impact noise is also important and Figure 9.4 shows

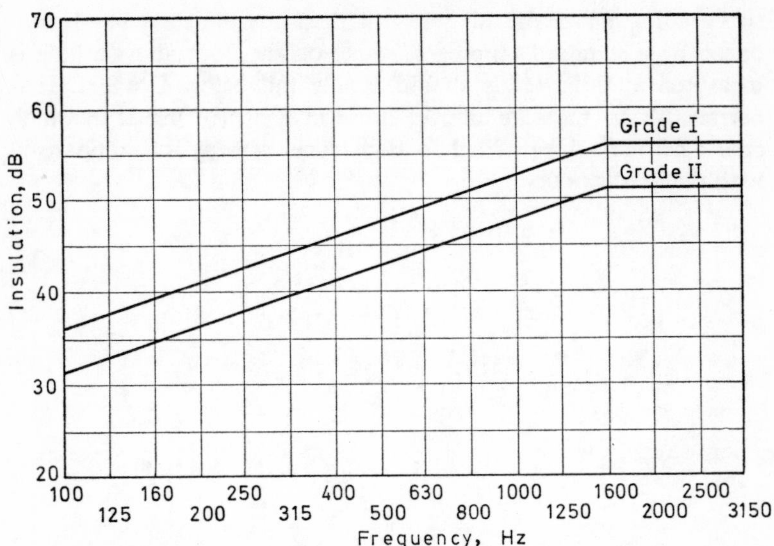

Figure 9.3. Recommended insulation between flats (U.K.).

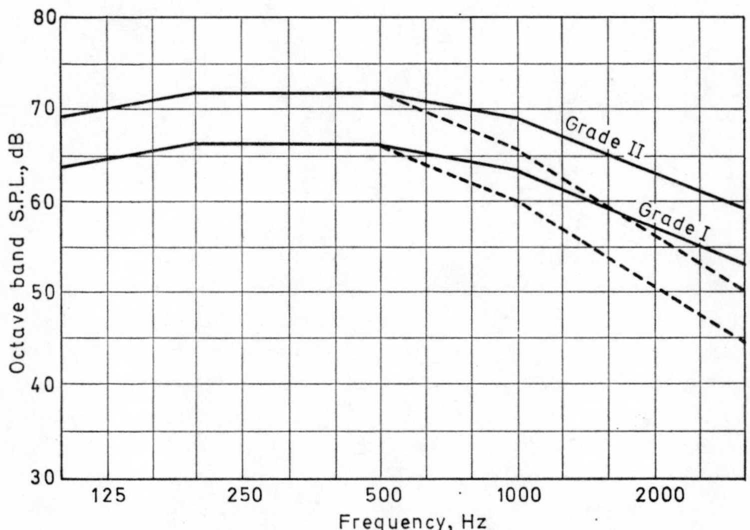

Figure 9.4. Recommended impact insulation between flats (U.K.): ——— measured without lino; . . . . . . measured with lino.

two grading curves. In this case it is the noise in octave bands produced by a standard tapping machine on the floor above which is measured and the levels should ideally fall below Grade I. It is permissible to measure impact noise in $\frac{1}{3}$ octave bands in which case 5 dB should be added to each measurement for comparison with the grading curve.

BIBLIOGRAPHY

L. L. Beranek, *Acoustics*, McGraw-Hill, 1954.

L. L. Beranek, *Noise Reduction*, McGraw Hill, 1960.

L. L. Beranek, *Music, Acoustics and Architecture*, Wiley, 1962.

P. V. Bruel, *Sound Insulation and Room Acoustics*, Chapman and Hall, 1951.

L. L. Doelle, *Acoustics in Architectural Design*, N.R.C., Canada, 1965.

W. Furrer, *Room and Building Acoustics and Noise Abatement*, Butterworths, 1964.

C. M. Harris, *Handbook of Noise Control*, McGraw Hill, 1957.

V. O. Knudsen and C. M. Harris, *Acoustical Designing in Architecture*, Wiley, 1950.

P. H. Parkin and H. R. Humphreys, *Acoustics, Noise and Buildings*, Faber and Faber, 1958.

E. G. Richardson, *Technical Aspects of Sound*, Elsevier, 1953.

H.M.S.O., Noise, Final Report—July 1963, Cmnd 2056, H.M.S.O.

C. M. Kosten and G. J. Van Os, *The Control of Noise*, N.P.L. Symposium No. 12, H.M.S.O., 1962.

British Standard Code of Practice, C.P.3, Chapter III 1960, Sound Insulation and Noise Reduction—B.S.I.

Insulation Handbook, Lomax, Erskine and Co. Annual.

Field Measurements of Sound Insulation Between Dwellings, National Building Studies Research Paper 33, H.M.S.O.

BS2750: 1956. Recommendations for Field and Laboratory Measurements of Airborne and Impact Sound Transmission in Buildings.

BS3638: 1963. Method for the Measurement of Sound Absorption Coefficients (I.S.O.) in a Reverberation Room.

London Noise Survey, H.M.S.O., 1968.

Handbook of Noise Measurement, General Radio Company 1963 (Available from Claude Lyons, Hoddesdon, Herts).

Noise from Motor Vehicles—Interim Report—July 1962, Cmnd 1780, H.M.S.O.

Traffic Noise, Publication No. 40, Greater London Council, 1966.

E. N. Bazley, *The Airborne Sound Insulation of Partitions*, H.M.S.O., 1966.

E. J. Evans and E. N. Bazley, *Sound Absorbing Materials*, H.M.S.O., 1960.

Building Research Station Digest No. 88 (1st series, Revised March, 1964).

INDEX